# The Environmental Impact Statement after Two Generations

This book is about a subject that Michael R. Greenberg has worked on and lived with for almost forty years. He was brought up in the south Bronx at a time when his neighborhood suffered from terrible air and noise pollution, and domestic waste went untreated into the Hudson River. For him, the National Environmental Policy Act (NEPA) was a blessing. It included an ethical position about the environment, and the law required some level of accountability in the form of an environmental impact statement (EIS).

Since then he has read the law and regulations that followed from it, worked on some EISs, read sections of many, and conversed with people who have prepared them and those who have reacted to them. And, while many analyses of the law and of the EIS process are helpful, they tend to be painted in black and white, without the subtleties and nuances that happen in the real world, and particularly away from the winner-takes-all forum of the law courts. Only a tiny minority of cases end up in the courts, and the vast majority of EIS processes are resolved by discussion and negotiation rather than litigation.

To properly evaluate an EIS means reading tables of data and lengthy case studies and illustrations as evidence, and reading environmental and risk evaluations that are painted in shades of gray – the analysts may have done a more than adequate job of measuring impacts on cultural artifacts, yet failed to assess the noise impacts adequately.

After forty years of thinking about and working with the NEPA and the EIS process, Michael Greenberg decided to conduct his own evaluation from the perspective of a person trained in science who focuses on environmental and environmental health policies. This book of carefully chosen real case studies goes beyond the familiar checklists of what to do, and shows students and practitioners alike what really happens during the creation and implementation of an EIS.

**Michael R. Greenberg** is a Professor and Dean at Rutgers University and the author of more than 20 books and 300 articles about environmental policy. He serves as Associate Editor for environmental health for the *American Journal of Public Health*, and is Editor-in-Chief of *Risk Analysis: An International Journal*.

# The Natural and Built Environment Series
Editor: Professor John Glasson, Oxford Brookes University

# The Environmental Impact Statement after Two Generations

## Managing environmental power

Michael R. Greenberg

Routledge
Taylor & Francis Group

LONDON AND NEW YORK

First published 2012
by Routledge
2 Park Square, Milton Park, Abingdon, Oxon OX14 4RN

Simultaneously published in the USA and Canada
by Routledge
711 Third Avenue, New York, NY 10017

*Routledge is an imprint of the Taylor & Francis Group, an informa business*

*British Library Cataloguing in Publication Data*
A catalogue record for this book is available from the British Library

*Library of Congress Cataloging-in-Publication Data*
Greenberg, Michael R.
   The environmental impact statement after two generations :
   managing environmental power / Michael Greenberg.
        p. cm. — (Natural and built environment series)
     Includes bibliographical references and index.
   1. Environmental impact statements—Law and legislation—United
   States. 2. Environmental impact statements—United States—Case
   studies. I. Title. II. Title: Managing environmental power.
   KF3775.G7184 2011
   344.7304′6—dc22                                          2011011108

ISBN: 978–0–415–60173–3 (hbk)
ISBN: 978–0–415–60174–0 (pbk)
ISBN: 978–0–203–80383–7 (ebk)

Typeset in Stone serif and Akzidenz Grotesk
by Keystroke, Station Road, Codsall, Wolverhampton

Printed and bound in India by Replika Press Pvt. Ltd.

# Dedication

∙∙∙∙∙∙∙∙∙∙∙∙∙∙∙∙∙∙∙∙∙∙∙∙∙∙∙∙∙∙∙∙∙∙∙∙∙∙∙∙∙∙∙∙∙∙∙∙∙∙∙∙∙∙∙∙∙

Faith Pappas, Max Wilkerson, Zane Wilkerson, Amelia Wilkerson, and Dylan Wilkerson, and billions of other children who are living with the decisions we make.

# Contents

# List of illustrations

## List of Tables

## List of Figures

# Preface

●●●●●●●●●●●●●●●●●●●●●●●●●●●●●●●●●●●●●●●●●●●●●●●●●●●●●●●●●●●●●●●●

I was born in the South Bronx in 1943, about a ten-minute walk from the old and new Yankee Stadiums. When I was seven years old, I had a bout with asthma, and as I grew older I developed serious pollen and some food allergies. Automobile traffic was growing, and filled the air with fumes that made my allergies worse. Our organic trash was burned in the apartment house incinerator, sending large clusters of ash through our apartment, if we (usually I) neglected to close the windows. Several nearby electricity-generating stations emitted foul-looking particles. When there was an air inversion in the fall, clouds of cough-inducing materials would hang in the air. The sewage from our area went untreated into the Hudson River, a fact I learned when I tried to fish in the East River. When I traveled to nearby coal mine areas in eastern Pennsylvania, I saw areas devoid of trees, enormous slag piles dumped in two adjacent valleys, and I could feel the heat from underground fires that had been burning for years. In short, before I was ten years old, I knew that we needed some strong government environmental actions; leaving environmental protection to the market meant more coughing and sneezing.

In 1969, I heard rumors that legislation was pending that would assert that the United States recognized the importance of environment. My first reaction was to not believe it. My second reaction was joy. When I read the law, my ambivalent reactions continued. I was impressed by the words, and yet the document offered little detail about implementation.

Forty years after the National Environmental Policy Act (NEPA) was passed, the ambivalence has not entirely disappeared. I have worked on some environmental impact statements, read sections of many, conversed with people who prepare them, and those who have reacted to them. I try to read books and regularly keep up with the *Environmental Impact Assessment Review*. And, when Congress decides to do a periodic review of NEPA, I read the reports. I'm not unhappy with the books, papers, and reports about NEPA. Yet, collectively, they have not "scratched my itch" about uses of the law. This is because evaluations of NEPA come across, whether intentionally or not, as polarized advocacy, not evaluation. For me, the arguments advanced are much too black-or-white and not sufficiently shades of gray. Many of the witnesses who have testified for or against the law appear to me to have been chosen because they advocate a particular position. And while their testimony is insightful, the evidence they present is too general for someone like me. I am used to reading tables of data and/or lengthy case studies and illustrations as evidence. I find the NEPA evaluation literature to be limited with regard to specific examples. I am used to reading environmental and risk evaluations that are painted in shades of gray. The worst EISs I have read still presented some useful information, and the best I have read have holes. When I talk about the EIS process in class, the students invariably, and I think appropriately, sit there waiting for me to go from generalizations to specifics. They want to know what was wrong with the noise impact analysis. When they search for examples, typically what they find are cases in which the courts played a major role. Evaluating the EIS process through court cases is like evaluating dentistry through implant or root canal surgery; only the rare case actually goes to court, and decisions from legal findings by definition are black or white.

In my personal experience, the EIS process is mostly a mundane planning process that generates a lot of information for decision-makers, some kernels of which are useful and thereby cause decision-makers to tweak their plans or dig their heels in and ignore the suggestions. I see the EIS process as a chameleon that changes form to suit the needs of the federal agency. I never thought that the US Department of Energy's EISs for nuclear weapons sites should be the same as the National Park Service's for a national park in the middle of the Great Plains. However, every EIS should be based on a consistent, multi-stage effort to obtain some consistent types of information, and this information should be presented in a way that is comprehensible to a reader without an advanced degree in science.

After forty years of puzzling about NEPA and the EIS process, I decided to conduct my own evaluation based on my understanding of NEPA, which has been strongly influenced by the literature and my own work. After thinking about the idea and doing initial designs, I identified three challenges. One issue is which EISs should be examined. The obvious solution is to conduct a random sample of projects. The problem with that approach is that I would need to have sufficient expertise on all the conceivable subject matter to understand the documents, which I do not have. Part of my solution was to interview at least one expert who was involved with each project. However, even the experts cannot remember all the details of a large EIS. Accordingly, I had to have sufficient expertise to understand the subject matter. Therefore the case studies I chose were those I felt sufficiently comfortable with.

A second challenge was evaluation criteria. My evaluations required strict adherence to a set of criteria. This is because there are so many environmental impact analyses about almost every conceivable subject. I needed to pick the most important criteria; ignore the idiosyncratic elements, unless somehow they truly were critical; and, most important, not get lost in every detail of every EIS. I settled on five standard questions about each EIS.

During my career, the vast majority of my publications were written for technical experts. This book is aimed primarily at students and their faculty teaching a class that is entirely, or has a section, on EISs. Students in environmental or civil engineering, environmental biology and chemistry, environmental planning and management, political science, environmental law, and other upper-level undergraduate and graduate courses are likely to be somewhat familiar with NEPA. However, they are unlikely to have any real examples that they can sink their teeth into, nor will they have exposure to how these cases exemplify, or fail to exemplify, what the creators of NEPA sought when the legislation and rules were written. I do not, however, want to exclude experts, such as some of my former students who have spent decades working on environmental impact assessments. Many of them tell me that they have become too specialized in one kind of assessment (for example, transportation impacts, or cultural or water quality analyses), and they also tell me they do not have the time to keep up with the ongoing assessment of the legislation. This book would allow them to broaden their understanding of the challenges to NEPA in the context of specific cases.

Michael Greenberg
February 18, 2011

# Acknowledgments

· · · · · · · · · · · · · · · · · · · · · · · · · · · · · · · · · · · · · · · · · · · · · · · · · · · · · · · · · · · ·

I begin by thanking some colleagues and organizations who directly helped with the book. Jennifer Rovito prepared the maps and selected the photos. Some of the photos are spectacular, showing the places much more effectively than we could with maps. I thank ESRI for having these available. I thank Mary Fetzer, a librarian at Rutgers University, for using her librarian skills to help me find a variety of EISs. Frank Popper, my colleague for more than twenty years, discussed the kinds of case studies to feature and was my expert for Chapter 7. I thank Kevin Brown, Governor Brendan Byrne, John Hnedak, Elizabeth Jeffery, David Kosson, Lauren O'Donnell, Frank Popper, Martin Robins, and Robert Waldman for talking to me about specific EISs. The vast majority of the maps are slight modifications of maps presented in the EISs acknowledged within the book. John Hnedak provided two photos from National Park Service archives shown in Chapter 3. We acknowledge those who assisted in preparing not only the maps and figures, but also the text.

I thank the leaders who sparked my interest in NEPA and the EIS. The 1970s started with an environmental melodrama. The first act was NEPA, starring President Nixon, Russell Train, William Ruckelshaus, Senators Edmund Muskie and Henry Jackson, and Lynton Caldwell as the key actors. This EIS requirement has lasted for four decades; while EISs are published, and Congressional hearings are held, the *Environmental Impact Assessment Review* is the major source of information that has sustained my interest, and I commend the journal's editors.

# 1 A statement of values and forty years of field trials

## Introduction

The National Environmental Policy Act (NEPA) became political necessity in United States when the sites, sounds, and odors of environmental damage became all too apparent to the American public. Highlighted by Rachel Carson's *Silent Spring* (1962) and Paul Ehrlich's *The Population Bomb* (1968), and by gruesome photos of devastation, pressure built for government inter-vention. Some argued that these books, along with magazine articles and photos, were overdramatic exaggerations of reality (Maddox 1972). Yet histori-ans and other scholars who have examined the impact of urban and industrial development during the nineteenth and twentieth centuries in the United States report major impacts on public health and on the water, air, and land environments. For example, Clay McShane (1994), David Stradling (1999), and John Cumbler (2005) paint national and regional portraits of the environ-mental insults of uncontrolled development and pollution. Andrew Hurley (1995), Martin Melosi (2001), and Joel Tarr (2003) focus on Gary (IN), the South, and Pittsburgh (PA), where industrial emissions and lack of infra-structure were particularly acute.

The stories in these fascinating books highlight a deep struggle between public health/environmental perspectives and the relentless, seemingly un-checked growth of capital, the inconsistent role of elected officials and govern-ment agencies in protecting public health and the environment, the exclusion of the most seriously affected populations, many instances of poor environ-mental science and distortion of science, and the lack of a moral imperative to protect the environment either for the current population or for future generations. These graphic American environmental histories were NEPA's context.

# NEPA: an overview

The National Environmental Policy Act (NEPA) of 1969 (Public Law 91-190) was signed by President Richard Nixon January 1, 1970, asserted a national environmental policy, and established the Council on Environmental Quality (CEQ). NEPA's preamble succinctly stated the law's broad objectives:

> The purposes of this Act are: To declare a national policy which will encourage productive and enjoyable harmony between man and his environment; to promote efforts which will prevent or eliminate damage to the environment and biosphere and stimulate the health and welfare of man; to enrich the understanding of the ecological systems and natural resources important to the Nation; and to establish a Council on Environmental Quality.
>
> NEPA, Purpose, Sec. 2, p. 1 (via www.epa.gov/lawsregs/laws/nepa.html).

Title I, the focus of this book, describes the policies and goals of the act; Title II created the CEQ, defined its responsibilities, and provided for its funding. The following section provides highlights of the act that are especially relevant to this book.

## Title I, Section 101

Section 101 has three parts. Part (a) declared that man and nature should exist in "productive harmony." Part (b) instructed the federal government to use its powers to achieve six objectives, as follows:

1 fulfill the responsibilities of each generation as trustee of the environment for succeeding generations;
2 assure for all Americans safe, healthful, productive, and aesthetically and culturally pleasing surroundings;
3 attain the widest range of beneficial uses of the environment without degradation, risk to health or safety, or other undesirable and unintended consequences;
4 preserve important historic, cultural, and natural aspects of our national heritage, and maintain, wherever possible, an environment which supports diversity, and variety of individual choice;
5 achieve a balance between population and resource use which will permit high standards of living and a wide sharing of life's amenities; and
6 enhance the quality of renewable resources and approach the maximum attainable recycling of depletable resources.

> NEPA, Title 1, Sec. 101, p. 1.

Title I, Section 101, part (c) reminds us that "each person has a responsibility to contribute to the preservation and enhancement of the environment."

## Title I, Section 102

Section 102 focuses on administration of law in eight parts (a–h). Part (a) tells federal agencies to use an interdisciplinary approach, including natural and social sciences, environmental design, and planning. Part (b) instructs the agencies to work with the CEQ and to include "unquantified environmental amenities and values [so that they] may be given appropriate consideration in decision making along with economic and technical considerations."

> Part (c) of Section 102 is the focus of this book. It calls for the "inclusion in every recommendation or report on proposals for legislation and other major federal actions significantly affecting the quality of the human environment, a detailed statement by the responsible official on –
>
> (i)   the environmental impact of the proposed action,
> (ii)  any adverse environmental effects which cannot be avoided should the proposal be implemented,
> (iii) alternatives to the proposed action,
> (iv) the relationship between local short-term uses of man's environment and the maintenance and enhancement of long-term productivity, and
> (v)  any irreversible and irretrievable commitments of resources which would be involved in the proposed action should it be implemented.
>
> NEPA, Title 1, Sec. 102, p. 2.

Part (c) also requires formal consultation. The responsible federal agencies are instructed to obtain comments from "any federal agency which has jurisdiction by law or special expertise with respect to any environmental impact involved," as well as "appropriate federal, state, and local agencies, which are authorized to develop and enforce environmental standards." Finally, a copy of the impact statement is to be made available to the President, the CEQ, and the public throughout the review process.

Provisions of sections d–i of 102 include, but are not limited to, recognizing environmental impacts as international, calling for efforts to support programs that are consistent with US foreign policy, making information about environmental quality available to states, counties and municipalities, as well as to institutions and individuals, and requiring responsible agencies to cooperate with the CEQ.

Section 103 of Title I requires the federal agencies to examine their policies, regulations, and authorities, and to propose to the President any adjustments that must be made to bring these into conformance with NEPA. Sections 104 and 105 state that NEPA should not contravene federal agencies' existing statutory obligations and that the NEPA's goals are "supplementary" to their existing authorizations.

# Initial interpretations

NEPA in general, and the environmental impact statement (EIS) specifically, is a legal tool to ensure that major federally initiated projects and new or substantially modified programs undergo a comprehensive review before committing major resources and beginning construction or implementation – in other words, it insists that federal agencies look before they leap, not that they cannot leap. NEPA calls for a multi-agency, multidisciplinary, open public assessment of the environmental impacts, as well as economic, health, and social impacts of each project's or program's proposals, and it requires consideration of alternatives to the proposed action.

The EIS forces consideration of NEPA's broad goals and preparation of an environmental impact assessment (EIA) on any proposed federal actions that could significantly affect the environment. It mandates that each EIS consider five broad environmental impact issues: environmental impact; unavoidable adverse impacts; alternatives to the proposed action; short-term uses versus long-term productivity; and irreversible commitments of resources. Responding to these five issues has been a challenge for federal agencies, which, with rare exceptions, previously had been able to design and implement projects with little if any consultation.

An assumption of the law is that intra- and inter-agency analysis, accompanied by input from private and public parties, will shape better decisions, that is, will avoid options that will exceed environmental standards and unduly burden populations, and at the same time promote options that enhance the ecological, economic, and social environments.

As noted above, NEPA was set up to be administered by the Council on Environmental Quality, which was established under Title II, Section 201. CEQ issued guidelines setting forth the rules and procedures agencies must follow to prepare an EIS. These guidelines (my copy from 1978 is forty-four pages long), with the aid of about forty years of practice and legal interpretation, have shaped the scope of the EIS document and its review.

The requirement to prepare an EIS applies to a long list of actions by federal agencies. The following six are illustrative, although they do include the bulk of conceivable projects (Kreske 1996; Bregman 1999):

- projects developed through federal grants
- planned federal projects
- legislative proposals
- changes in agency policies and operating procedures
- actions requiring federal licenses, permits, and other approvals
- actions with possibly controversial impacts.

The requirements of Section 102 (2) (c) have led to inclusion of the following six topics in EISs:

- description of the existing environment
- description of alternatives
- probable impacts of each alternative
- identification of the alternative chosen and the evaluation that led to this choice
- detailed analysis of the probable impacts of the proposal
- description of the techniques intended to minimize any adverse impacts.

Court rulings have demonstrated that federal agencies cannot leave out any information. EISs have also been rejected because credible alternatives to the proposed action were not considered at all, or alternatives were not adequately presented; this includes the option not to undertake any project at all. The beneficial or adverse impacts of each alternative are expected to be estimated by comparing the expected future conditions of the preferred alternative with existing conditions, the no-action alternative, and any other alternatives presented in the EIS. Hence, ideally, a multidisciplinary group should design and carry out an EIS, because a multidisciplinary team is unlikely to overlook an important impact.

A full EIS is the well known option for responding to the requirements of part (c) of Section 102. But it is not the only option. When a federal agency proposes an action, it has three options, as follows:

- "Categorical exclusion" is an option if the federal agency and CEQ decide that the action does not pose a threat individually or collectively to the environment. Typical examples would be road and bridge repair – in essence, minor adjustments. Categorical exclusion has become more controversial in recent years (see Chapter 8).
- An environmental assessment (EA) is the first step to determine if a full-blown EIS is required. If the EA process finds no significant impact, then a "finding of no significant impact" (FONSI) is made. However, if the proposal implies some "extraordinary circumstance," then a full-blown EIS can be required. There may be some public involvement in the EA process. (See Chapter 7 for an illustration.)
- An EIS is required for proposals with the potential to substantially impact the environment.

The process has seven steps:

1  The EIS process begins with a notice of intent to prepare an EIS.
2  This is followed by a scoping process, in which the investigation is designed. Public involvement is critical at the scoping stage (see Chapter 2 for an illustration). A good scoping process should prevent serious stakeholder-related problems later on.
3  The third stage is preparation of a draft EIS. Each agency has its own protocol, but each would go through the following steps. The agency would

list the information needed, and review and then categorize the adequacy of available data. It would begin filling critical data gaps and decide on a format for a draft EIS. Then it would prepare a project work plan and begin to fill in all the data gaps to prepare a draft EIS. Various parts of the preliminary draft EIS would be prepared as the information is gathered and analyzed.

4   Sections would be circulated to internal staff and modified. After reviewing all internal staff comments, the federal agency would prepare a draft EIS for distribution to other federal agencies, other government bodies, and the public.

5   These external parties would review the draft EIS and comment, doubtless leading to some revisions of the document, albeit not necessarily alteration of the proposed action recommendation.

6   A final EIS is then prepared, circulated and presented.

7   After reviews have been completed, including responding to comments, the federal agency issues a record of decision.

Federal agencies differ in the extent to which they use categorical exclusion, adopt an environmental assessment as the final document, or proceed to a full EIS. For example, over a two-year period, about 90% of proposed Federal Highway Administration projects used categorical exclusion, 7% environmental assessments, and 3% EISs. The Army Corps of Engineers produced about 4500 environmental assessments and only seventy-five to a hundred EISs. The Department of Energy prepared forty-nine environmental assessments and eleven EISs (Committee on Resources 2006).

The CEQ and the US Environmental Protection Agency (EPA) are responsible for reviewing the EISs. The CEQ determines if the federal agency has met the law's requirements. The EPA is charged with evaluating the adequacy of the documents. The EPA can declare a draft EIS "adequate," "insufficient information," or "inadequate." Insufficient information has come to mean that more information is required to assess the impact and/or that plausible alternatives have not been adequately assessed. A declaration of inadequate implies a need for substantial data gathering and analysis.

Tzoumis and Finegold (2000) and Tzoumis (2007) have examined trends in EPA ratings of the quality of draft federal EISs, beginning in the 1970s. The authors observed that many EISs do not have a formal rating. With that caveat noted, 64% of draft EISs were rated by the EPA as insufficient information, 32% as adequate, and 5% as inadequate information. The authors expected the rating of draft EISs to improve, but that proved not to be the case.

At the final EIS stage, the EPA's options range from no objection to three levels of concern about the proposed action. A declaration of "environmental concern" indicates that the EPA has identified actions that should be avoided in order to "fully protect" the environment. "Environmental objections" is a stronger declaration, pointing to impact that should be avoided in order to even "adequately" protect the environment. The strongest objection is "environmentally unsatisfactory," which expresses opposition to the proposed alternative because of expected serious adverse environmental impacts. Tzoumis and Finegold (2000) and Tzoumis (2007) report that a little over 1% of EISs were rated as "unsatisfactory," and 51% had a rating of "no objections." The remaining 48% received ratings with some objections.

Objections to EISs are common – these typically range from issues concerning information and science in the case of government agencies to disagreements about values in the case of citizen groups (see Chapters 2, 4–5, and 7 for illustrations). After a quarter-century of challenges, the federal government created the US Institute for Environmental Conflict and Resolution as part of the Morris Udall Foundation, to try to resolve differences and to train individuals as dispute-resolution professionals. The Institute brings people together, develops a plan to resolve the issues, selects a facilitator, oversees the process, and in many other ways will try to settle conflicts before they are brought to court. However, the Foundation notes that success is not guaranteed, and it describes some of the reasons why a negotiated settlement sometimes is not possible (www.udall.gov). Its recommendations are consistent with this author's experiences, and Chapters 4 and 7 illustrate two cases where the value issues are quite striking.

The federal court system is the ultimate decision-maker, and the US courts have been a key player in defining each agency's role and obligations under NEPA (CEQ 1997a,b, 2007; Weiland 1997; Committee on Resources 2006).

While the role of the federal courts is often highlighted in debates about when and how to apply NEPA, in reality some of the federal agencies have developed sophisticated protocols for determining which actions require an environmental assessment or EIS (see for example US Department of Energy 2009a; also see CEQ 1997a; Committee on Resources 2006). Overall, after four decades of trial and error, the environmental impact process has evolved into a broadly applied and emulated legal device. What did its designers hope for, and what were initial experiences?

## Recalling NEPA's architects

NEPA was shaped by elected officials and their advisers, who saw environmental protection as a major public policy issue. Perlstein (2008) suggests that President Nixon did not care much about environmental protection one way or the other (p. 460); he was in search of a constituency (see also Flippen 2000; Lindstrom and Smith 2001). However, Nixon's colleagues John Ehrlichman (former land-use attorney) and General Curtis LeMay were self described environmental supporters. In addition, Senators Edmund Muskie (known by some as "ecology Ed" and Henry "Scoop" Jackson were seen by Nixon and each other as likely opponents in the 1972 presidential election. Nixon was aware that environmental concern among the public had tripled since 1965.

The President made a pragmatic decision to support environmental protection, with NEPA as the first clear legislative signal. President Nixon, asserts Flippen (2000), believed that he could not win the 1972 presidential election with an environmental platform, but he could be beaten up without one. Senator Jackson focused a good deal of his attention on NEPA, whereas Senator Muskie focused on efforts that would lead to the Clean Air Act and the Clean Water Act.

Stepping back from the 1960s political environment that created NEPA, the law has been widely praised for stating a national environmental policy, an environmental ethic and vision of the future, and for emphasizing multi-disciplinary thinking and planning (CEQ 1997a,b; Kaufman 1997; Committee on Resources 2006). Kaufman (1997) declared it to be the USA's sustainable development policy before the expression was popular. In 1969, Senator Jackson of the State of Washington, the chief legislative architect of NEPA, believed it to be the "most important" and "far reaching" environmental and conservation measure ever enacted in the United States, and he hoped that the CEQ would advise the President of the United States and that the EPA would work to protect environmental quality by its regulatory powers (Alfano, undated).

The original version of NEPA did not include the action-forcing mechanism of Section 102 (2) (c). Professor Lynton Caldwell of Indiana University, who was a consultant to Senator Jackson, developed the idea for an action-forcing mechanism. He called for federal agencies to evaluate the effect of their proposals on the environment, and he emphasized the need for these analyses to be thoughtful and rigorous (Caldwell 1982, 1989). Caldwell and other proponents felt that the impact statement was a strategic counter to possible indifference or even hostility toward NEPA by federal agencies, elected legislators, and President Nixon, who had an uncomfortable relationship with the environmental community.

While Section 102 forced an action by the federal agencies, Dreyfus and Ingram (1976) struggled with some of its less-than-precise terminology. For example, the expression "to the full extent possible" can be interpreted as requiring a substantial amount of information, or only information that is relatively easy to acquire. That impact statements were to be prepared for "major federal actions significantly affecting the quality of human environment" is also open to debate. Is a major federal action construction of a new span over a major river? Or is it merely repaving the bridge and adding better lighting? Perhaps the most debated element is the call for describing "alternatives" to the proposed action. Are these alternatives all variations of the same theme, or are they entirely different approaches?

Dreyfus and Ingram (1976) were forgiving of NEPA's drafters because they realized that the law was going to be applied across many different kinds of projects, by many different agencies. Professor Caldwell (1989, 1998) demonstrates his ambivalence when on the one hand he labeled NEPA as the American environmental "Magna Carta," and on the other hand notes that it has not fulfilled its potential because of inconsistent application by federal agencies, narrow definition by the federal courts, and wide variations in interest in the law by the executive branch.

While Caldwell (1989, 1998) could not ignore the limitations of his creation, and called for a constitutional amendment to elevate environmental protection to be equal to property and civil rights, Dreyfus and Ingram (1976) concluded that NEPA had already achieved far more than anyone could have anticipated in the United States and internationally. They focused on widespread access to the NEPA process and the legal leverage of participants (see also Keysar 2005).

The most obvious ambivalence generated by NEPA that this author has read is the assessment of Oliver Houck, who gives Caldwell more credit than Caldwell gives himself. First, he says that:

> At the end of the day, NEPA did not do all that it was intended to do. Its goals remain goals. Its impact statement mechanism has indeed become something of the catechism, an institutionalized ritual in which the original words have lost their meaning.
>
> (Houck 2000, pp. 178–180)

And yet he adds:

> NEPA's great contribution . . . is the environmental impact statement. It is not what the statement says that is important. It is in what comes before, and what agencies have to investigate and learn and listen to, and what they have to fear from other agencies and environmental groups, the press, the reviewing court and in the everyday responses and accommodations they have to make. The NEPA ideas of disclosure, public participation, alternatives, and judicial review are blockbuster stuff as well as for the developed countries of Europe and are absolutely revolutionary stuff for developing nations in Latin America in the Far East and for those, like Croatia and Cuba, who are also signing on. In this one regard, NEPA has been the largest environmental success in the world.
>
> (*ibid.*)

## Initial media attention

NEPA was signed on January 1, 1970, and controversies underscored by the ambivalence in these introductory comments began immediately. I use four news articles to summarize the tone. Writing in *The New York Times*, Hill (1970) characterized the initial report of the CEQ as "less a record of accomplishment than a laundry-list of problems to be grappled with." While praising the idea of moderating "the environmentally traumatic propensities of tunneled-visioned Federal agencies," he characterized the first five months of the program as "rough" and noted that so far only twenty-five impact statements had been done, and none on the supersonic transport (SST).

In November, *The New York Times* (1970) reported that: "the Nixon Administration appears to have decided that it can withhold environmental impact studies from the public until the decisions they influence have been made and announced." The article quoted CEQ chair Russell Train as follows: "there is no question that there is a tendency to prepare a section 102 statement after a decision has been made." The article provided examples of when federal agencies released impact statements at the time of making the decision, and examples of impact statements released in advance of their decision.

A few days later, Senator Philip Hart of Michigan stated that "several executive agencies are undermining the effectiveness of the National Environmental Policy Act by failing to file required environmental impact statements" (Kenworthy 1970a). A spokesperson for the Council reported that they had insufficient staff to review even the 200 EISs that they had received.

On December 2, 1970, Senator Muskie accused the administration of not making available critiques of the SST prior to the congressional vote on it. A spokesperson for the US Department of Transportation replied that it was inappropriate to release reports in a piecemeal fashion (Kenworthy 1970b). With the SST as the focus of their ire, congressional representatives continued to assail the administration for suppressing criticism of the SST. Senator William Proxmire of Wisconsin, for example, stated "if the National Environmental Policy Act is to have any teeth at all, Congress and the public must have access to comments by Federal Agencies on proposed Federal actions, whether or not those comments happen to support the program under consideration." (Kenworthy 1970c). The administration was accused of "government secrecy of the worst kind, making a mockery of the democratic process and of the National Environmental Policy Act."

The year 1971 was a much better year for President Nixon's environmental initiatives and NEPA. Nixon introduced a broad set of environmental proposals (Flippen 2000; Perlstein 2008) that led to the clean air and water legislation. With regard to NEPA, Russell Train (1971), chair of CEQ, announced that CEQ proposed new federal procedures for EISs that would require that the public receive and be able to make comments in the early stages of EIS preparation. William Ruckelshaus, first head of the EPA, was praised for his candor and a variety of actions, including a letter criticizing the Alaskan oil pipeline EIS (Kenworthy 1971a). The most obvious change in 1971 was that organizations such as the Natural Resources Defense Council, Environmental Defense, and the Sierra Club began to use the EIS process to confront federal agency decision-making about coal mining (Vecsey 1971), dams (Bryant 1971), underground nuclear testing (Lapp 1971), the "big sky" resort south of Bozeman, Montana (Kenworthy 1971b), drilling for oil near Santa Barbara, California (*New York Times* 1971a), the Teton River dam project in Idaho (Blair 1971a), the Alaskan pipeline (*New York Times* 1971b), coal leases west of the Mississippi River (Franklin 1971), a nuclear power plant in Illinois (*New York Times* 1971c), a hydroelectric facility (Kenworthy 1971b), a highway extension on southern Long Island (Andelman 1971), offshore oil leases (Blair 1971b), and others. These cases involved actions proposed by the Tennessee Valley Authority, US Army Corps of Engineers, US Atomic Energy Commission, US Forest Service, Department of the Interior, and Department of Transportation. In each of these cases, the media used reviews of EISs by EPA, national environmental organizations and local citizens, and elected officials.

From no public information about environmental impact, within a year of the passage of NEPA, thousands of new environmental policy documents were being created. Notably, on June 12, 1971, the federal government announced that the National Technical Information Service (NTIS) would provide a subscription service, summarizing EISs and other federal environmental action, for five dollars a year. NEPA had created a new product to market, and the author subscribed to it for many years, and periodically ordered a microfiche copy of an EIS.

## Praise and criticism

NEPA and its state and local progeny have had four decades of field trials. This section summarizes the major strengths and then weaknesses of the EIS process.

### Praise

Beginning with strengths, one important accomplishment of this process at the federal, state, and local levels has been its widespread use as a substitute for planning. The EIS has become in practice an explicit assertion that development requires early planning to avoid degrading the quality of the environment (Best 1972; see also Chapter 8). Early assessments typically emphasized the potential for better agency decisions. For example, Caldwell in a foreword to a book by Richard Andrews (1976, pp. xi–xiii) observed that the federal departments and agencies had become too specialized, unable to see past their specific missions, and insulated from public values and attitudes. Caldwell described NEPA as an example of American self-government.

NEPA is often viewed as a conscious-raising statement, expressed locally in the EIS and nationally in programmatic EIS analyses. The law rejects the idea of economic growth without consideration of environmental, public health, and social consequences, and without reflection. An EIS should offer a more comprehensive, interdisciplinary analysis, including estimates of the secondary and induced effects of a proposed action, and should stimulate an appreciation of the complexity of large-scale developments.

The EIS process created a system of checks and balances between federal, state, and local governments. A federal agency can legally ignore opposition to a proposal by other federal agencies, states, localities, and the public. But this is a dangerous political step, as government agencies typically try to avoid open clashes with their counterparts (Oregon State University 1973). Best (1972, p. 19) recognized the challenge to federal agencies, noting that "the most significant consequence of the new policy is that it places the burden of proof on the initiators of new developments – the federal agencies and their clients in industry."

Part of NEPA's appeal is the widespread assertion that it has changed practice by federal agencies, which has led to better decisions. NEPA, say its proponents, has been instrumental in the cancellation or postponement of highways, dams, airports, nuclear waste disposal programs, outer continental shelf leases, and other proposals. More often, the scoping, preparation, and presentation of the results have caused changes in locations, designs, and other changes to mitigate undesirable environmental effects (CEQ 1997; Committee on Resources 2006). Doubtless, there are many such instances, and it would be fruitful to have these documented systematically (see Chapters 2 and 7 for examples).

While NEPA's gentle nudge upon inter-federal policy outcomes merits more acknowledgment, impact on other government bodies is almost always acknowledged. NEPA's scoping process is critical insofar as it directly pits the tendency to be multidisciplinary, all-inclusive, and precautionary to protect public health and the environment, brought to the table by agencies that are not proposing a specific project, against the tendency of the agency proposing the project to try to economize by narrowly construing the scope of the EIS (Snell and Cowell 2006, Jain *et al.* 2002). NEPA clearly encourages multi-disciplinary agency and community participation, including obtaining input and requiring carrying through comments from multiple parties through the process. NEPA's call for public participation has been well received (see Chapter 8 for further discussion). Early public participation should reduce opposition.

In the original act, the federal government is the focus. However, some states and local governments in the United States and other countries have developed NEPA-like processes. Their requirements vary. Compared with the national law, the progeny typically require less detailed content analyses, and focus on private actions as well as public ones (there are exceptions, such as California). They include a wide variety of administrative or overseeing approaches, pre-emption with regard to NEPA (federal law will normally take precedence), and wide differences in public participation requirements. Writing a few years after the passage of NEPA, Hagman (1974) evaluated some of the strengths and weaknesses of these progeny, although he noted wide variation by state. His praise focuses on NEPA primarily as a substitute for land-use planning in states that have very little power to control land use, and on the efforts he sees by states in their legislation to balance environmental, economic, and social concerns. His major concerns are the fear that aesthetic criteria that protect the interests of affluent individuals will be used to route unwanted land uses through poor areas, and that a state EIS is a "ridiculous waste of resources in comparison to a comprehensive planning process" (Hagman 1974, p. 48). Chapter 8 discusses this author's views of NEPA as a substitute for a real US planning legislation.

Most local government EIS requirements are directed primarily at private development, and the responsible overseers are usually planning, zoning, and environmental commissions. Some of the overseers have the legal authority to approve, deny, or request the alteration of a development proposal, where others are advisory.

Yet across these different geographic scales – international, regional, state, and local – EIS staff have cited instances ranging from the preservation of dunes and clusters of unusual trees to the cancellation and modification of massive projects. Cumulatively, these cases led some early commentators to conclude that the EIS process has been successful (Oregon State University 1973).

To summarize the praise, the EIS process should lead to better planning; raise people's consciousness; result in better communication between scientists, agency leadership, and the public; and create a series of checks and balances, especially when there is good leadership (Keysar and Steinemann, 2002). Andrews (1976) characterized NEPA at the time of passage as "unremarkable." It had only two titles, could be printed on four pages (my original is five pages), and was uncontroversial, passing through Congress in ten months, with no dissenting votes in the Senate and only fifteen in the House. Andrews added that, in the first few years (1970–75), many proposed federal actions had been modified because of NEPA, and also that twenty-two foreign nations (now well over 100) and some states and local governments had emulated NEPA.

## Criticism

The criticisms of NEPA probably exceed the praise, at least as this author measures them by volume of pages collected. Some have charged that the EIS process has been less effective than it could be. One repeated criticism is the inherent contradiction in NEPA and many of the state and local progeny. The law creates the opportunity for considering environmental factors in government actions. Yet Section 102 (2) (c) is a procedural requirement. If the procedure is followed correctly, the federal agency can make a decision that many think severely degrades the environment (Steinemann, 2001). Jenny (quoted in Oregon State University 1973, p. 14), a former member of CEQ, characterized CEQ's role as a "soft-voiced advisor and commentator, and where necessary, guide." He went on to assert that Congress wanted the federal agencies to take this responsibility upon themselves, and not for the CEQ to be an enforcer.

Agencies can arguably circumvent the intent of the law by not securing involvement until the proposal is too far along to be changed, although it is more difficult now than forty years ago. In other words, an excellent document may make no difference. Frustration with this problem led Fairfax (1978) to argue that NEPA has wasted environmentalists' resources on paper shuffling, and that their time would have been more effectively spent trying to change agency decision-making authority.

A second criticism of NEPA and some state laws is that private sector development is not included. Since the vast majority of development is by the private sector, this should be a serious shortcoming, although some state and local governments have developed NEPA-like requirements for private and not-for-profit projects, and more important private development is included if federal grants are part of a project.

Administrative discretion is a third issue. Some contend that many actions with environmentally significant impacts are not accompanied by an EIS because agencies decide that the actions are not "major" or "significant," or do not constitute an agency "proposal" or "action." In addition, to avoid preparation of an EIS, or to make sure that one is not vulnerable to legal opposition, documents are infused with as much information as possible to protect the agency's position. Although the information is sometimes interesting, it often amounts to a conglomeration of not exactly relevant information that avoids the significant environmental impacts of a proposal. This problem is especially evident in the discussions of alternatives, which have been criticized for being limited and perfunctory by some, and unrealistically narrow by others.

Yet another criticism focuses on scientific evaluation of impacts. No matter how detailed or comprehensive the document may be, and even if the EIS is scoped by an interdisciplinary team, there is no way of ensuring that all impacts considered significant by all parties will be included. Some scientific facts and relationships are missing because the team did not think about them; a good argument for their inclusion was not offered to the project design team; or information or data were lacking.

Lawrence (2000a,b,c) notes that not all impacts are significant, and accordingly not all missing information is important. He offers suggestions for defining a significant impact. Atkinson *et al.* (2006) assess court responses to missing information, and report that courts have compared the need for missing information with the cost of acquiring it, and have considered the possible adverse effects of taking action without the data. To avoid a delay, Atkinson *et al.* (2006) recommend that agencies should explain how the benefits of receiving are greater than the costs of delay.

Two further concerns related to science are measuring and weighting impacts. The impact of some environmental hazards is not known, that is, some have multiplicative rather than additive effects on people and the environment. Another challenge is comparing the impacts of clearly measurable effects, such as dissolved oxygen and pH, with more subjective indicators of aesthetic and cultural impacts. At best, researchers are comparing apples (e.g. water quality, air quality) and oranges (e.g. cultural artifacts, job affects)

(Oregon State University, 1973; Ortolano 1973; Warner 1973; Baecher *et al.* 1975; Cheremisinoff and Morresi 1977; Greenberg *et al.* 1978). Overall, the limits inherent in estimating environmental impacts are an unending source of criticism of the EIS. It is always possible to cast suspicion on the science of impact statements. And because some view the EIS as an advocacy document rather than a scientific document prepared to engage stakeholders, they are inherently suspicious of the data and the conclusions (Oregon State University 1973). The case study chapters in this book (2–7) offer ample opportunities for such questions.

Some experts feel that so-called "evangelistic" environmentalists and some members of the public become too polarized and refuse to grasp the scientific subtleties considered in an EIS (Oregon State University 1973). The counter is that instead of focusing on the important impacts, some assert, experts try their best to avoid providing scientific "ammunition" to adamant opponents, rather than isolating and dealing with the key environmental problems.

Person (2006) offers a strong argument that the most difficult sticking points are about values and interests, not about data and models. Confronting values and interests is more important than trying to gather more and better information. Vicente and Partidario (2006) argue that the success of the teaching environmental assessment depends upon enhancing communication among a myriad of stakeholders with different beliefs, convictions, values, experiences, needs, and other factors, all of which lead to different world views (see Chapter 8). Wood (2008) offers recommendations for communicating the significance of environmental impacts, but primarily explains how difficult it is to present scientific results while also placing them in context.

The most widely publicized criticisms from both proponents and opponents are economic. Some argue that government has made little investment in the EIS process. They cite substantial staff cuts at CEQ and EPA to argue that more resources are needed to protect and preserve the quality of the human environment. They add that the EIS process has created employment for environmental and social scientists, administrators, and lawyers. Without the EIS process, arguably some of the major advancements in monitoring the environment and modeling contaminants would not have been made. These studies, argue proponents of the EIS, were more than worth the effort because they will avoid both short-term and multigenerational impacts that could have imposed greater future costs. However, one of the standard criticisms of EIA is its failure to examine and portray cumulative impacts, including economic ones (CEQ 1997; Committee on Resources 2006; Tang *et al.* 2009; Warnback and Hilding-Rydevik 2009). It is one that is voiced in many of the case study chapters in this book.

In contrast, the loudest voices during the past fifteen years have focused on the economic costs associated with the EIS process. Representatives of the business community contend that the EIS process is a slow-growth policy, and in some cases a no-growth policy, that hurts the economy. They point to time involved in preparing and reviewing EISs and the inflationary impact on costs.

This inflationary impact is further increased when delays result from NEPA litigation and potential injunctions, which they underscore in their arguments.

## Two post-year-2000 assessments

This section focuses on two of the many reviews of NEPA, conducted during the administration of Presidents William J. Clinton and George W. Bush. I have chosen these two because they clearly demarcate different views of the EIS process prior to the Obama administration.

### NEPA: Lessons Learned and Next Steps. Oversight Hearing. *US House of Representatives, 109th Congress, First Session (Committee on Resources 2006)*

I have quoted liberally from the report because the specific words of some of these witnesses are illuminating, and even some of the more diplomatically worded testimony is refreshingly frank.

One hearing was held in Washington DC and five others at sites across the nation, with a clear southern and western geographic bias: Spokane (WA), Lakeside (AZ), Nacogdoches (TX), Rio Rancho (NM), and Norfolk (VA). The Committee invited representatives with different starting perspectives. The spokespersons provided verbal and written testimony, and were questioned by members of Congress. Several of the invited guests provided supplementary testimony in response to questions from members of Congress. Several members of Congress expressed their views. This record is so important because the unedited testimony sharpened the language of the witnesses.

The eighty-eight-page transcript does not call for the complete abrogation of the law. Every witness supported the statement of environmental principles. Representative and Chairperson Cathy McMorris from the State of Washington summarized her views as follows: "We have heard countless times and in countless ways that NEPA is a good law born of good intentions." (p. 1). James Connaughton, chair of CEQ under President George W. Bush, noted: "NEPA is remarkable for its simplicity. [It is] a story of innovation, [its] fundamental objectives are as relevant today as when Congress passed it" (p. 2). He added that NEPA has stood the test of time and is a policy for future generations of Americans. Yet he did not wait very long to question the implementation of the program, noting that CEQ regulations that implement NEPA are only twenty pages long (see Nicholas Yost testimony below) and need to be "fine-tuned" for specific action, such as forest management and energy development (see Chapter 4). Connaughton noted considerable conflict among federal agencies over NEPA and the need to address timeliness, increase effectiveness, and reduce intergovernmental conflict.

In response, Congressman Tom Udall from New Mexico retorted (p. 15):

It now seems clear that my view of NEPA differs significantly from the views of those who have come before the Task Force to criticize the statute. Where they see delay, I see deliberation. Where they see postponed profits, I see public input. Where they see frivolous litigation, I see citizens requiring the government to live up to its responsibilities. And where they see a barrier to development, I see a shield that protects average Americans from the shortsightedness of a massive Federal bureaucracy.

Following Congressman's Udall's testimony, Congressman Nick Rahall of West Virginia added a less diplomatic statement of his own about the committee investigation (p. 17):

. . . the record developed by this Task Force is extensive, but it is not sufficient. To the extent that this Task Force was designed to provide a body of evidence for the need to amend NEPA, it has failed. Perhaps this is because this Task Force plowed very little new ground. Both the Clinton and current Bush Administration conducted comprehensive reviews of the law and made specific recommendations for improving implementation without amending the statute.

Or perhaps it is because the argument forth by industry witnesses – that federal agencies should act less deliberately and enable more rapid public lands profiteering – failed to resonate with an American public stung in the wallet by huge energy conglomerates, likely the greatest beneficiaries of NEPA changes that are now enjoying the largest profit in American history. Or perhaps it is because the credibility of this Task Force was repeatedly undercut by this Committee when it made sweeping changes to NEPA in the energy and budget reconciliation bills despite the fact that this Task Force had not completed its work.

Next to testify was Nicholas Yost, an attorney who was the general counsel for CEQ during President Carter's administration. Yost, in fact, took the lead in drafting the CEQ and NEPA regulations. After noting that NEPA has had only one amendment in twenty-five years, he provided his perspective (p. 20):

I strongly believe that NEPA and its basic message, look before you leap environmentally, served the American people immensely well. This statute has been environmental success story. It's been replicated in about half of our States, and it has served as a model for environmental impact assessment laws in more than 100 countries and may be the most imitated law in American history. Also I should point out that I have spent much of the last 20 years assisting clients through the NEPA process and have had my own share of frustration with unneeded delay in the process. The goal should be to cut the fat but not the muscle.

Yost argued that actions should not be exempted from NEPA, consideration of alternatives should not be reduced, the public should be kept in the process, and judicial review is essential. While offering constructive suggestions and criticisms, Yost provided historical context for the current hearing (p. 22):

> I respectfully suggest that we keep in mind the original intent of the drafters. The Senate's lead author, Henry Jackson of Washington, characterized NEPA as "the most important and far-reaching environmental and conservation measure ever enacted." The ranking Republican, Gordon Allott of Colorado, called it "truly landmark legislation." The lead House author, Congressman John Dingell of Michigan, stressed that "we must consider the natural environment as a whole and assess its quality continuously if we wish to make strides in improving and preserving it." President Nixon chose January 1, 1970, to sign NEPA into law as his first official act of the new decade.

Next to testify was Robert Dreher, Deputy Executive Director of the Georgetown Environmental Law and Policy Institute. He echoed much of Ned Yost's testimony, and added some insights based on his empirical evaluation of NEPA. Dreher was Deputy General Counsel to the EPA and staff attorney to the Sierra Club Legal Defense Fund. He characterized the Task Force's record as "oddly limited" (pp. 39–40):

> Unfortunately, the Task Force to date has focused on a narrow, and almost uniformly negative, set of concerns: complaints raised by representatives of businesses that use federal public lands and natural resources for economic benefit that compliance with the Act's procedures imposes burdens and delays on their activities. The Task Force has shown little apparent interest in how NEPA protects environmental values, in fulfillment of Congress's original goals for the Act. Perhaps for that reason, the Task Force appears not to have been particularly interested in the views of conservationists and recreationists who, not surprisingly, see the value of NEPA. . . . Moreover, the Task Force virtually ignored the people with the most hands on experience in implementing NEPA: federal officials responsible for complying with the Act. . . . My report, NEPA Under Siege, describes these assaults on the act, ranging from measures in the 2003 Healthy Forests Restoration Act that restrict analysis of alternatives and limit public participation in forest thinning projects to the rebuttable presumption established by the Energy Policy Act of 2005 that numerous oil and gas activities are categorically excluded from NEPA analysis. Cumulatively, these proposals threaten to kill the NEPA process with a thousand cuts.

Quoting liberally from the Congressional Record and former President Nixon, Dreher reiterated the mantra that NEPA does not force agencies to accept recommendations, but it does require the opportunity for presentation of options. Presenting empirical research (notably not seen elsewhere in these hearings), Dreher disputed the assertion that NEPA is a major cost element or a source of delay, and in fact he argued that other actions related to business and government are the major source of inefficiencies. With regard to NEPA-related law suits, he provided data to show that few law suits are filed and very few injunctions are granted by the courts; in other words, NEPA is not a major source of legal activity to block projects.

Yost and Dreher were followed by a number of witnesses with a business perspective. John Martin, an attorney for the Devon Energy Corporation, identified the issue of increasing magnitude of EIS requirements. Rather than the 150–300-page documents anticipated by the regulations, he reported documents that are many thousands of pages long, cost millions of dollars, and require two to three years to prepare. The environmental assessments he noted have grown from ten to fifteen pages to hundreds. Martin advocated a reduction in the number of alternatives that must be considered; public participation early in the process so that plans can be adjusted while it is relatively simple to do; and a statute of limitations of 180 days placed on challenges so that opponents cannot wait until the last minute to sabotage a project.

Nick Goldstein, a staff attorney for the American Road and Transportation Builders Association, argued that "in its current state, NEPA generates far more documents than decisions" (p. 48). By delaying road projects, he argues, we waste dollars, thwart local planning efforts to manage land use and reduce congestion, and allow more auto-related injuries and deaths.

## The National Environmental Policy Act, A Study of its Effectiveness after Twenty-Five Years (CEQ 1997a)

NEPA is described as the "foundation of modern American environmental protection" (p. 1) and later as an "eloquent and inspiring declaration" (p. iv). According to the study's authors, NEPA has "made agencies take a hard look at potential environmental consequences of their action, and . . . has brought the public into the agency decision-making process like no other statute, . . . and as a tool it has helped to 'build community and to strengthen our democracy" (p. iii).

The criticisms are similar to those in the later 2005 document, but are toned down. Businesses asserted that NEPA produces documents to satisfy a requirement, not to improve decisions; that it costs too much; and that the regulations are at odds with other federal laws, causing confusion, delays and increasing costs, and leading to unnecessary law suits. Their pro-NEPA counterparts counter that agencies wait too long to consult; the public believes that its concerns are ignored and that cumulative effects are ignored by many EISs. Advocates noted that about 500 EISs are completed a year, compared with 50,000 environmental assessments. They assert that more EISs should be done, and that agencies are thwarting the intent of the law by relying so much on environmental assessments.

The most valuable element of the CEQ report is the group's reflection on five elements of the NEPA, as follows:

- Strategic planning – NEPA values should be integrated into agency planning while it is possible to modify plans.
- Public information and input – the public should be informed and heard.
- Inter-agency coordination – agencies should share information and plan jointly.
- Interdisciplinary place-based analysis and decision-making – information and values from place-based sources should be included.
- Science-based and flexible management – predictions should be monitored and evaluated.

## My decades-old thesis, now revised

When I began learning about NEPA, I understood, or at least hoped, it to be a science-driven reform mechanism. The early literature up to about 1985 (especially Best 1972; Oregon State University 1973; Ortolano 1973; Warner 1973; Andrews 1976; Dreyfus and Ingram 1976; Liroff 1976; Canter 1977; Greenberg et al. 1978; Caldwell 1982; Wenner 1982; Taylor 1984) suggested that the EIS might work as such, in four possible ways.

- It would introduce a rational process into the federal agencies, especially emphasizing ecological effects that agency decisions insufficiently weighed.
- It would produce comprehensive systems analyses of the proposed action's ecological, human health, economic, and social impacts, and thus those of projects, and thereby move the government toward more thoughtful project decisions.
- It would inject young and enthusiastic federal employees, such as many of my former students, who would internally reform agency thinking about the importance of the environment.
- It would open up federal decisions to local publics and activist groups, and lead to a broadening of the factors considered by decision-makers. This would also take into account the social tensions that arouse in cities during the 1960s.

But by the mid-1980s, I had worked on and read EISs, spoken with experienced EIS practitioners, and become more informed by reality than by my values. While not entirely ready to abandon the science-driven, rational model of decision-making, my none-too-surprising conclusion is that agencies make decisions through leaders with little time to leave their mark. To them, an EIS that supports the agency's position will be acceptable. One that requires minor changes will be tolerable – indeed, an EIS that leads to tweaking the preferred option is good, because it shows that the agency is listening to in-house analysts and outside audiences. But comments that strenuously critique the agency's goals are ignored, if possible. The federal courts might rule that an EIS contains mistakes that require correction, but there have been few such procedural mistakes recently. In a few cases, an EIS could change an agency's mind if it found serious consequences supported by irrefutable evidence. But most of the time, by the mid-1980s I expected that short-term, mission-based decision-making would continue to dominate.

I also expected to find wide variations in agencies' development and use of the EIS because of their markedly different missions. Hence I expected to find more informal process and open negotiations around a mass transit project or a monument preservation project, because the projects are desired by most residents. By contrast, the siting and operations of a liquefied natural gas or a nuclear power plant, usually opposed by nearby residents, would lead to a more formal, less open process.

I also expected that the EIS, whether the agency loves it or hates it, is often its *de facto* planning process. Many American leaders and ordinary citizens do not like planning (Popper 1993) because it prevents bold actions, has little public support, and often degenerates into legalistic paper chases. The EIS planning process not only requires agencies to follow a sequence of predetermined steps, but also forces them to consider issues that they otherwise would have ignored or preferred to avoid. My overall expectation prior to doing the case studies was that the federal EIS has become an environmental chameleon that fits each agency's needs to manage its environmental power and its internal planning processes (see Chapter 8 for further discussion).

## Evaluation questions

Given my revised thesis about the EIS, I developed a set of specific questions for the evaluation. These were derived from three sources. One was the intent of the creators of the law, as summarized above. Second, I reflected on the critiques of NEPA, some of which are also summarized above. Third, I constructed the questions so that they incorporated a set of six policy-evaluation criteria that I developed and have used in environmental health policy analysis for many years (Greenberg 2008). My experience has shown that environmental policies that fail to respond adequately to the following six factors have a very high probability of failing. These six criteria are:

- the likely reaction of elected government officials and their staff
- likely reactions from the public and special interests, including not-for-profit organizations, business, and the media
- human and ecological health impacts
- short- and long-term economic costs and benefits

- the moral imperative
- flexibility and time pressure.

It follows, then, that NEPA should allow for explicit consideration of advantages and disadvantages of proposed actions from these six perspectives. For example, as I am writing this chapter, there is considerable debate about the desirability of closing the prison at Guantánamo Bay and moving the detainees elsewhere. Assume that the United States was going to build another high-security prison for the special detainees (some already exist that could host many of them). An EIS for such a facility would need strong support from the Departments of Defense, Homeland Security, and Justice. Presumably the EPA and the Department of the Interior would also have input, depending on the location. States and local areas selected as possible locations should have an opportunity to provide input during the EIS process. Special interests, both public and private, as well as media, should be included during the scoping and draft EIS phases. Depending on the location, ecological issues, cultural artifacts, and human health could be major considerations. The cost of designing, building, and operating the facility for an indeterminate length of time would need to be estimated and be part of the EIS. With regard to flexibility and time pressure, important considerations would include the consequences of not opening the facility in the immediate future. In other words, what options are available in the immediate future that would allow the decision to be deferred or more consideration given to it during the next five years? Ethics and morality might not be explicitly labeled as such in an EIS, but surely the international implications of Guantánamo and local implications of hosting such a prison would be considered. If any of these six criteria were missing from an EIS process, the proposed action would be likely to have serious problems.

I began with the idea of using a fourth source for key questions – a comparison of the US NEPA with its foreign and US state and local progeny. NEPA is lauded, by even its critics, because over 100 countries and over twenty-five US states have some NEPA-like process. But emulating does not mean copying. I examined the EIA processes of China, India, Mexico, and Canada, as well as those of California, New Jersey, Maryland, Virginia, and Minnesota, and the cities of San Francisco and New York. The more I tried to draw lessons learned from these, the more I realized that I was comparing apples and oranges. There are good reasons why other countries, states, and local governments might have a different EIA process from that of the United States as a whole.

China requires every EIS practitioner to be licensed, which I personally think is a good idea. Yet China's government has been much more centralized than that of the United States, so that difference is not surprising. I like the fact that San Francisco used its EIS processes to manage housing and other land uses and to prevent environmental injustice. Yet, while the United States does have the Department of Housing and Urban Development, that department has been substantially weakened and almost dissolved. Management of housing typically has been left to states, especially local governments, in the United States. I also like the European Union's requirement for private as well as public project EIAs. However, some US states and local governments also require private developers to do EIAs, and EU nations differ in the extent to which this is required and enforced. Furthermore, the literature shows that each of these nations, states, and local governments has EIA management problems that derive from these "improvements" (Ruddy and Hilty 2008). Overall, after pondering international, state, and local issues, I decided that it would be foolish to try to include these perspectives in this book because I cannot possibly do justice to them here.

I believe it is presumptuous to say that the framers of NEPA should have included in the legislation or regulation "modern" concepts such as strategic environmental assessment, life cycle analysis, and life cycle cost analyses, although I understand and have used these concepts. NEPA was crafted to assess environmental, public health, economic, social, and cumulative effects of each proposed action. A programmatic EIS is a reasonable extension that can address cumulative impacts and could be expected in some of the older EISs. The more recent ways of looking at the environment and other differences between the NEPA and its progeny are worth exploring, but are too much of a leap of faith for purposes of what is largely a retrospective analysis of forty years of experience in the United States (Partidario 2000; Therival 2004; Chaker *et al.* 2006; Fischer 2007). A number of these options are discussed briefly in Chapter 8.

With these three sources acknowledged, I focused on five multi-part questions for each case study in this book.

- Does the EIS report offer an adequate and objective analysis of the information? This implies an evaluation of the quality and quantity of the data, the analysis of the results, and the presentation, including its tone.
- Does the EIS include integrated environmental, economic, and social considerations? This means a review to determine if the authors were comprehensive and emphasized key factors and cumulative effects, if any.
- Is there evidence in the document of meaningful coordination with other federal, state, and local agencies? This necessitates a review of comments and actions by other government agencies.
- Was the process accessible to nongovernmental organizations, citizens, and the media in ways that gave them an opportunity to provide meaningful input? As above, this requires an evaluation of meaningful access to the process.
- What would have happened to this project if there had not been an EIS process?

## Organization of the book

The book is divided into eight chapters. Each of the case study chapters (2–7) follows the same format, beginning with an introduction that places the case study in its geographic and temporal context. The second part of each case study discusses the proposed action(s) and the final decision. Then the history of the proposal, its design, conflict, and resolution are reviewed. The penultimate part of each chapter is an interview with one or more experts. The final section of chapters 2–7 is my evaluation of the EIS, and in one case the EIA, process. The choice of case studies evolved over the course of about four years and included discussions with colleagues who had many suggestions. Ultimately, I chose to feature a diversity of agencies, including examples from every major region, cities and rural locations, some early 1970s and some newer EISs, and notably a final EIS, draft EIS, and EIA, and a scoping exercise for an EIS. Applying these criteria led me to about two dozen choices, and then I fell back to choosing from cases that I felt were within my scientific understanding and for which I could identify experts who were willing to speak with me.

This chapter has described the history of the EIS in the United States and reviewed the five questions asked about every EIS in this book. Chapter 2 looks at the issues of transportation, sprawl, and urban revitalization in New Jersey, the most densely populated state in the United States, using two EISs – one about a major highway project that was defeated; the second about a light rail system that is partly completed. Chapter 3 illustrates how an EIS may be used to preserve cultural/historical treasures (Ellis Island in the New York Bay). Chapter 4 is about how EISs are done when hazardous substances are involved, in this case a proposed liquefied natural gas facility and distribution system (near Baltimore, Maryland). The fifth chapter uses an older EIS to examine the destruction of a small portion of the US chemical warfare agent stockpile (Johnston Island, Pacific Ocean). Chapter 6 is about the challenging task of managing high-level nuclear waste at one of the former US nuclear weapons facilities (Aiken, South Carolina). Chapter 7 is an environmental assessment about what might be the last major dam project in the United States, Lake Nighthorse, near Durango, Colorado.

These six case chapters vary in length in response to the scope of the project being studied. A limited number of maps, pictures, and tables are provided for context. The smallest of these EISs was about 250 pages, and the largest was thousands of pages long. While I list all the major topics considered in the EIS, I deliberately chose key elements to emphasize in the text, those that are most important based on my personal knowledge and public comments. Also, I think it is essential that readers see for themselves that EISs are written in markedly different styles, that have different tonal qualities when read, which depend upon the mission and culture of the organization preparing the EIS. This will become apparent as the reader finishes reading Chapter 3 (Ellis Island) and begins reading Chapter 4 (Sparrows Point liquefied natural gas). Chapter 8 concludes the book with my summary judgments about the five questions asked of each case study, and my views on how the EIS/EIA process can be modified to improve its utility.

# 2 Metropolitan New Jersey: transportation, sprawl and urban revitalization

······································································

## Introduction

The end of the Second World War released pent-up demand for new housing in the United States. Fishman (2000) describes how government housing programs, new highways, and other government programs promoted suburban development. Deindustrialization and then civil unrest of the 1960s increased middle-class suburbanization. Post-war transportation increasingly became automobile-oriented. Rates of car ownership doubled, or more, in major American cities such as Boston, Denver, Detroit, Los Angeles, San Francisco, and even mass-transit-oriented New York City. Automobile use increased even more in the newly formed auto-dependent suburbs (Roberts 1999).

Between 1950 and 1970, California, Florida, and New York experienced the largest relative population increases, and some of their older industrial neighborhoods declined. Yet the author believes that New Jersey was the poster child for the negative impacts of auto-oriented suburbanization and industrial erosion. With a population of 1200 people per square mile, New Jersey has the highest population density. Indeed, it is the only state in the United States with a population density exceeding 1000 per square mile. Between 1950 and 1970, the "Garden State's" population increased by 2.3 million people, the sixth largest proportional increase of any state in the United States. Suburban Burlington (Philadelphia suburb) and Middlesex, Monmouth, and Ocean counties (New York City suburbs) grew by more than 40%. Ocean County's population (see Figure 2.1) increased 93% between 1960 and 1970, the largest increase of any metropolitan county in the United States during that decade.

Compounding the increase of commuter automobile traffic in this densely populated region, many residents of New Jersey, New York, and Pennsylvania

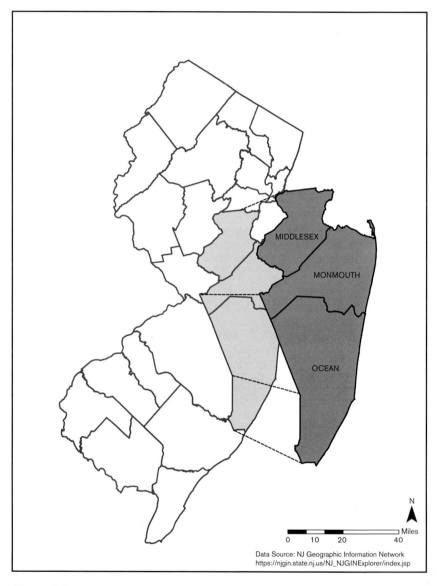

**Figure 2.1**

Locations of Middlesex, Monmouth, and Ocean Counties, New Jersey

have taken their summer vacations along the New Jersey Atlantic Ocean shore (see Figure 2.1). Driving to the New Jersey shore from New York City, Philadelphia, their suburbs, and northern New Jersey became a time-consuming and grinding experience.

After the Second World War, New Jersey had the highest proportion of its labor force in manufacturing jobs. But as these jobs gradually disappeared from cities, Hudson and Essex counties (see Figure 2.2, Essex is located directly west of the Hudson), two densely developed, industrial-oriented counties lost population, industrial, and commercial jobs. Many new roads were proposed as essential to revitalizing these cities.

In short, during the late 1950s and 1960s, building more roads and widening existing ones seemed like a straightforward solution to managing increasing traffic congestion and revitalizing cities. Yet, by the 1970s, the undesirable environmental consequences of the road building remedy became obvious. Oxides of nitrogen and photochemical smog, produced by automobile engines, increased. The national ambient air-quality standard for ozone was exceeded in many places in the United States, more so in New Jersey than other states; nineteen of the twenty-one New Jersey counties exceeded the standard. During the summer, a brown haze formed and drifted hundreds of miles from the Philadelphia suburbs through New Jersey and New York State into Massachusetts, Vermont, and elsewhere in New England. The smog hung over farms, forests, and other rural areas, frustrating many people and violating the newly promulgated national air-quality standards.

Highway-associated site and sound problems became increasingly obvious. New highways brought disruptive noise to formerly quiet neighborhoods. Some viable communities were destroyed by new roads and widening of roads. People accustomed to walking to visit their friends a few blocks away were separated by six-lane roads. Open space in cities, as well as in suburbs, was lost, and the toll on animals, plants, and ecosystems grew. Most distressing to transportation planners was that new highways did not seem to reduce traffic. Instead, more highways seemed to generate even more traffic.

In a study for the state of New Jersey, Burchell *et al.* (2002) have documented the economic and psychosocial costs of unmanaged sprawl experienced by California, Florida, and many other states, including and perhaps especially New Jersey. One cost was that more investment was required in new suburban areas for schools, police, fire, water and sewer systems, and other services. While new schools were being built in suburbs, schools in many cities and older suburbs deteriorated and were closed when their population declined. Hence the second part of the economic penalty of unmanaged suburbanization was the increasing abandonment of sunken investments in cities and older suburbs. Burchell *et al.* (2000) estimated that a continuation of the suburban-oriented recent land-use trends in the state would require additional annual infrastructure and service costs of approximately $418 million (about $55 per capita). By instituting moderate efforts to control suburbanization, the additional annual deficit would be reduced to $257 million, or $160 million less.

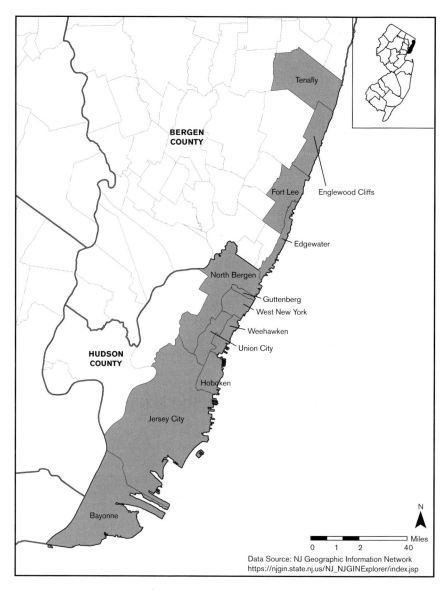

**Figure 2.2**

New Jersey's "Gold Coast"

Focusing on two New Jersey transportation projects, this chapter considers the impacts of rapid post-Second World War suburbanization and urban decline, especially the environmental impacts. More specifically, I examine how the environmental impact process intersected with policies, first to defeat more highways, and then to choose among mass transit proposals. The two

EIS processes presented here represent uses of the EIS process to manage transportation-related environmental decisions; and, secondarily and strikingly, the adaptation of the EIS process to the realities of the late twentieth and early twenty-first centuries. The reader will see two documents that are remarkably different, especially with regard to public participation.

New Jersey offers numerous illustrations of the rejection of highway proposals. I found forty-six highway projects in New Jersey that were canceled or modified between the early 1970s and the early twenty-first century. Nearly all of these rejections occurred before 1980. They were located in every county, and a number crossed into New York and Pennsylvania. They ranged from small, unfinished segments of existing highways to new 50-plus-mile four- and six-lane roads. The three major reasons I could find for their defeat were increasing costs, local public opposition, and environmental requirements. From among these failed proposals, I selected the Alfred E. Driscoll Expressway, which was bitterly contested, with environmental concerns at the forefront of the debate. With regard to mass transit, proposals for light rail systems, monorails, and other mass transit people-movers have blossomed. Three light rail systems were constructed in New Jersey: Hudson–Bergen, Newark, and the River line. While each offered interesting environmental and political challenges, I picked the Hudson–Bergen Light Rail, which involved lengthy debates.

I believe that the EIA process highlighted the problems associated with highway projects and the advantages of mass transit. The New Jersey Department of Transportation stopped at least six of the forty-six failed projects because of the need to prepare an EIS. State officials concluded that serious environmental impacts could not be avoided. For example, some proposed routes would have passed through the environmentally sensitive Pine Barrens in southern New Jersey, now a large protected ecological preserve with dwarf pine trees and other uncommon ecosystems. Other projects would have passed through the New Jersey Meadowlands, another unusual ecological system, a part of which contains the football stadium of the Giants and Jets. Other highway proposals would have encroached on the Passaic Falls, where Alexander Hamilton planned the first American manufacturing center. And still other projects would have required building a new bridge across the Hudson River on the majestic Palisades scarp that is visible from the George Washington Bridge.

I am not saying that no new highways were built after NEPA was passed; rather that there clearly was a major scaling back of highway projects after the passage of NEPA. Increasing public concern about the environment made highway projects politically vulnerable when their EISs were scrutinized. Elected officials and citizen groups quickly recognized that they could use the NEPA requirement to make their case for preserving open space and historical facilities, and for pressing both anti-sprawl and highway agendas.

It was feasible to design engineering solutions to some of these environmental problems, but the engineered solutions made some highway projects too expensive. Also, some of the proposed projects defeated in New Jersey would have been blatant examples of environmental injustice, destroying city neighborhoods occupied largely by poor, minority populations. Although the expression "environmental injustice" did not become part of the common lexicon until the late 1980s, critiques of highway proposals showed how they disproportionately impacted poor neighborhoods. In short, the EIS process clearly helped inform decision-making, providing information to support arguments against many highway proposals.

A good deal of the public, and many powerful elected officials, turned against building more roads and increasingly toward mass transit. In 1972, New Jersey's voters defeated a $650 million transportation bond issue that emphasized highway construction. And again, in 1974 and 1975, a combination of commuter groups and environmental organizations worked actively to defeat highway-oriented transportation bond issues in New Jersey (Sullivan 1979). When Brendan Byrne became Governor of New Jersey in 1974, his platform emphasized improvements in mass transportation and opposition to highways.

Among the numerous road and rail projects in New Jersey, I have picked two that illustrate the role of the EIS in highlighting the abrupt change in public perception of new highways, and the interesting role of the EIS in the process of underscoring this change.

## The Alfred E. Driscoll Expressway: proposal and decision

### History of the proposal

In 1964, the New Jersey Highway Authority, the then operators of the Garden State Parkway, proposed a 45-mile-long expressway, the Garden State Thruway, linking Woodbridge in Middlesex County, New Jersey with Toms River in Ocean County (see Figure 2.1). This thruway, to be open to all vehicles, including trucks, was to provide a bypass of the Garden State Parkway (no trucks allowed on the northern part of the highway) and US9 (slow traffic with many traffic lights) through the interior of the state. Along with an east–west highway, this northwest-to-southeast toll road was part of the so-called "Central Jersey Expressway System" between New York City and Philadelphia. The New Jersey Highway Authority already had purchased 253 acres along the proposed route. Plans for the Garden State Thruway were on official maps through much of the 1970s. However, noting financial cost concerns, the New Jersey Highway Authority decided against building the Garden State Thruway.

In October 1970, the New Jersey Turnpike Authority took over the project and changed it in several important respects. In 1971, the Authority proposed a 36-mile-long, four-lane toll expressway from the New Jersey Turnpike in South Brunswick (Middlesex County) to the Garden State Parkway in Toms

River. The name was to be the Alfred E. Driscoll Expressway, after the governor who opened the New Jersey Turnpike.

The Driscoll Expressway was to provide a high-speed corridor for central and southern New Jersey, and notably was to provide access to trucks. The $1.2 billion expressway (dollars inflated to the year 2010 by author) was to be financed by New Jersey Turnpike Authority bonds. The Driscoll Expressway was to open in 1976 and to carry approximately 40,000 vehicles per day (annual average daily traffic, AADT). While its construction would have required the displacement of eighty-four homes and six businesses (see EIS discussion below for details), the expressway (in conjunction with appropriate planning and land-use controls) was expected to accommodate residential, commercial, and light industrial growth more effectively than the existing set of highways. Lapolla and Suszka's (2005) book about the New Jersey Turnpike characterized the road as "an extension of the Turnpike's beautification program" (p. 99). Readers who have driven along the New Jersey Turnpike would be hard pressed to describe it as "beautiful." Hence characterizing the new road as an effort to extend beautification, frankly, lacks credibility. I will try to put this observation in perspective in the following section.

## Design concepts

Architects and engineers had learned from negative reactions to the New Jersey Turnpike's long and straight lanes and lack of any semblance of aesthetic values that the public wanted something more visually appealing. Hence, while the plan had some of the corridor-like efficiency of the Turnpike, the Alfred E. Driscoll Expressway was designed to include aesthetic values that were consistent with the increasing interest in environmental protection of the early 1970s. For example, the Driscoll Expressway was to have a median that preserved existing vegetation. Landscaping along rights-of-way and center medians was to reduce noise from the road and make it appear greener and less gray. A greenbelt along the Expressway's 450-foot-wide right-of-way to enhance the environment reinforced this objective. In addition, the road was to have controlled access at interchanges spaced miles apart. Interchanges were to be constructed at seven locations along the 36-mile road (Figure 2.3).

The highway was to have 12-foot-wide traffic lanes, 12-foot-wide shoulders, and 1200-foot-long acceleration and deceleration lanes. All grades were to be kept to a maximum of 3% to increase safety and improve appearance, and all curves were to have a minimum radius of 3000 feet. The expressway was designed for 70 mph, yet the legal speed limit of 60 mph was established to allow for a margin of safety. In addition to these design standards for the roadway, there were also specifications established for all turnpike structures, including bridges and storm-drainage facilities. In other words, compared with the exceeding linear, undecorated, and dreary-looking New Jersey Turnpike, the Driscoll Expressway was to be less visually offensive.

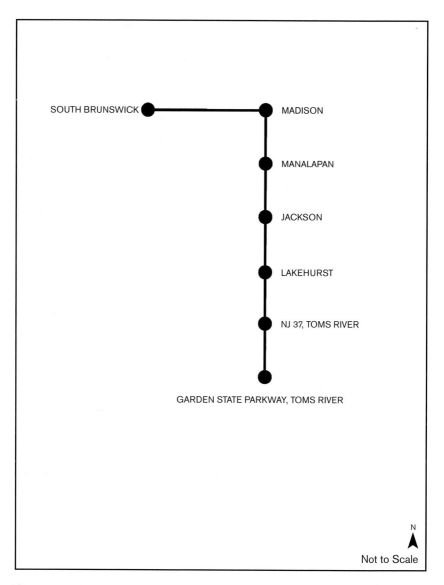

SOUTH BRUNSWICK ●━━━━━━━━━━● MADISON

● MANALAPAN

● JACKSON

● LAKEHURST

● NJ 37, TOMS RIVER

●

GARDEN STATE PARKWAY, TOMS RIVER

N
▲
Not to Scale

**Figure 2.3**

Exits on proposed highway

## The environmental impact statement

In May 1972, the New Jersey Turnpike Authority hired a team of four expert consulting groups to evaluate the potential environmental impacts of the proposed Driscoll Expressway. New Jersey's guidelines for EISs called for the experts to "provide the information needed to evaluate the effects of a proposed project on the environment" (State of New Jersey 1972; *New York Times* 1973a). In September 1972, the team submitted a 250+ page report (New Jersey Turnpike Authority 1972). That EIS, consistent with that period, is relatively short and not equivocal about issues.

Before highlighting some of the details, note that the document's conclusion is that the "completed expressway will provide a pleasing, functional and economically sound addition to the southeastern part of New Jersey" (*ibid.*, p. i.) The authors add:

> The need for additional access to serve commercial, passenger and mass transit traffic has been apparent for many years. Intolerable congestion, stop-and-go traffic and unconscionable delays have been too typical of May–October travel. This situation now extends to virtually an all-year travel pattern.
>
> (*ibid.*, p. ii)

The firm of Howard, Needles, Tammen & Bergendoff described the project and reviewed alternatives, including the no-action alternative, any probable adverse environmental effects that could not be avoided, and steps to minimize environmental effects. They noted some visual aesthetic impacts of the proposed road, especially during construction. They also mentioned rights-of-way and fences, but that these would be obscured by landscaping. They lauded the idea of a green central median. They pointed to no major positive or negative impacts on man-made resources, nor did they identify negative impacts on the health, safety, or well-being of the public. They pointed out that seventy-three homes and eleven mobile homes would be displaced, along with six active businesses. With regard to health and safety, they noted that the Expressway was designed to minimize long, straight stretches of the roadway, and that rest areas would be provided for driving breaks, as well as emergency services to assist injured people and disabled cars. The consultant concluded that accident rate would be much lower than for the roads currently carrying the traffic to that part of New Jersey, in fact, about a quarter of the rate for traffic headed in the same direction on existing roads.

Bolt, Beranek and Newman, Inc. evaluated noise impacts. They estimated that a total of twenty-five residences, including ten mobile homes and fifteen scattered houses, would be impacted, and that 15% of three parks and recreation areas would be affected by the road traffic. The company pointed to schools, including Manalapan High School, which would be affected, but noted that noise barriers would be built to eliminate the impacts. Notably, they emphasized that noise levels in the area as a whole would be reduced by, on average, 4–5 decibels (dBA) by transferring traffic, especially trucks, from existing roads to the Driscoll Expressway. In my opinion, this was the best known noise-impact consulting firm in the nation at this time.

Coverdale & Colpitts, Inc. reviewed socioeconomic impacts. Their review is remarkable for its long list of positive impacts and no negative ones. For example, the proposed road would relieve summer peak-hour congestion, provide access to trucks, improve highway safety by diverting vehicles from existing roads, decrease vehicle-operating costs, improve prospects for better commuter-bus services, increase land values and thereby tax bases, and attract new light industry and commercial facilities as well as new residential growth. Their evaluations suggest that more land would be committed to permanent open green space and that, while a small portion of viable farms might be affected, many other farms would be converted to higher-value end uses. Any negative impacts, they concluded, could be managed by applying local zoning and other ordinances and by proper enforcement of regulations.

Environmental Research & Technology, assisted by other firms, examined air quality, water resources, other natural resources, and social and physical resource impacts. No adverse environmental impacts on air quality were noted, nor any on water resources. The authors point out that the route requires several rivers to be crossed and that some parks would be crossed, but that these crossings, and wetland fillings, would be done without damage by following proper practices. Invoking a "two wrongs make a right" argument, they noted that the processes used for the highway would be less destructive than those for other developments in the area. They did identify potential impacts on cranberry bogs in the state, and once again emphasized that these problems could be prevented by proper design and construction.

With regard to social and physical resources, their report shows no major adverse environmental impacts; rather, they stated that impacts could be addressed by design and construction. They pointed to positive impacts on some public facilities that would become more accessible and defined. They also described why some route alternatives were not recommended, almost always citing incompatibility with local plans.

The no-project plan was described as follows:

A decision not to build the Governor Alfred E. Driscoll Expressway would mean a continuance of the existing inconvenience to local highway users, an increased . . . inconvenience to vacationers traveling to south shore towns, an intensification of traffic congestion and the ignoring of master plans prepared by affected municipalities and counties.

. . . A decision not to construct the Alfred E. Driscoll Expressway would waste the efforts of the affected municipalities and counties to shape desirable future regional development patterns, protect existing and potential county parks, road and correctional facilities and protect local patterns of development.

(*ibid.*, pp. 17–18)

As noted above, this document in particular, like many of that early period, did not devote a great deal of space to equivocation.

## Conflict and rejection

The Driscoll Expressway was approved by the New Jersey State Legislature in 1972, and the EIS was prepared and submitted in September 1972 (see below). Public hearings began shortly thereafter along the proposed route, and sixteen alternatives were considered.

Local community groups raised major objections during these hearings. Their concerns were about noise, added traffic, possible air pollution, and harm to the New Jersey Pine Barrens, a unique ecological system in central and southern parts of the state. For example, on November 26, 1972, residents of suburban Manalapan Township criticized the proposal, focusing on the design that indicated it would pass within 200 feet of an elementary school and would require removing twenty-three homes (Cheslow 1972). On December 15, a meeting in Toms River brought out some proponents but many more opponents (*New York Times* 1972). Again, this kind of public reaction to early EISs was common.

Yet the Turnpike Authority did not yield. It countered the opposition, and Wall Street immediately purchased $210 million of the bonds within a few days to fund the highway (Dawson 1973). In September 1973, the Turnpike Authority met with mayors of four municipalities who pressed for modifications in the alignment (*New York Times* 1973a). In November, the Turnpike Authority awarded the first contract for the expressway (*New York Times* 1973b).

But less than a month later, Governor Brendan Byrne opposed the highway as a "danger to the environment and [because] the fuel shortage had reduced the need for the road" (Sullivan 1973). Yet, having spent $20 million in design and engineering, a day later the Turnpike Authority contested the Governor elect's opposition (Waggoner 1973a). Former governor Alfred E. Driscoll, chair of the New Jersey Turnpike Authority, asserted that former Governor William T. Cahill had approved the highway and that his successor did not have the authority to reverse that decision. But, on December 28, 1973, plans to award any additional contracts were postponed (Waggoner 1973b).

Six months later, a three-judge appeals panel supported the opposition, criticizing the authority for misleading the public about the route (Sullivan 1974). Governor Byrne was criticized by construction workers (Janson 1974) and investors (Phalon 1975), and was asked not to attend the funeral of former governor Driscoll (Sullivan 1975). A week after governor Driscoll died in March 1975, the New Jersey Turnpike Authority officially dropped its plans for the Driscoll Expressway. During the late 1980s, the Expressway rights-of-way were sold.

# Interview with former governor Brendan Byrne

Brendan Byrne was Governor of New Jersey from 1974 to 1982. I had spoken to him about various environmental and public health issues during the years. Consequently, when I called him on July 17, 2009, he immediately understood the purpose of my book and got right to the point. He made two major points. The former governor recognized the legal and political significance of NEPA, especially in the 1970s, when the Driscoll Expressway was proposed, and he supported the idea of comprehensively analyzing environmental and social impacts. Yet his reaction to this proposal, which I quote below, illustrates that asking the right questions does not necessarily lead to answers that decision-makers support, especially when the EIS, frankly, does not seem to consider the full range of viewpoints and data. Governor Byrne said:

> I made a political decision based on gut feeling and advice. A lot of what a governor does is based on a gut feeling. I concluded that the Expressway would've led to the [New Jersey] shore being swamped by developers and tourists. One local official, who was against, said that the Expressway would have led to the paving over of Ocean County. State Senator John Russo, who has strong environmental credentials, supported my position at the time. The Expressway was not a major political issue. The EIS did not convince any of us to support the Expressway. I just was against it, and I had good reasons for my opposition. I never regretted the decision.

Governor Brendan Byrne, I believe, because of his political sophistication and understanding of the entire state, had a more nuanced and I would say accurate understanding of the cumulative impacts of this proposed expressway than the supporters of the highway and those who prepared the EIS. If a different person had been Governor of New Jersey at that time, there was a good chance that the Driscoll Expressway would have been built. Governor Byrne did what most elected officials will not do, which is to overrule a high-priority project of a cabinet-level agency (see Chapter 7 for another example).

# Evaluation of the five questions

## Information

The information presented in the document appears to this author to be based on appropriate science and engineering. For example, the noise experts were arguably from the best noise-pollution company in the United States, and they identified places where decibel levels would cause public reaction. From reading the EIS, I cannot see a single location where the noise analysis was improperly done. Yet there are several inexplicable holes in the information, most notably the impact on open space. Although I admit I have not surveyed the area, the open space analysis seems perfunctory, but I may be wrong. Unfortunately, given the period when this document was prepared, I cannot locate any documents or people who can testify to the thoroughness of the open space analysis.

The most serious problem is the judgmental tone of the document, which must have offended anyone who might legitimately have a different opinion. Indeed, I imagine that the tone of this document inflamed the passions of opponents and swayed people who might be neutral toward opposition. For example, the experts who worked on the EIS concluded that regional traffic congestion, noise, and air pollution would be reduced by the Driscoll Expressway, and that local problems would be controlled by engineering. Residents did not assume that noise barriers would be built, many did not want noise barriers, and many, perhaps most, simply did not want any latent intrusive sounds. They attacked the interpretations of the data on the quality of their lives, not the science *per se*.

## Comprehensiveness

The document includes environmental, economic, and social considerations, emphasizing the importance of land-use and transportation planning. It looks at cumulative local effects and even regional effects, and asserts that all of these will benefit as a result of the Driscoll Expressway. What is missing is any sensibility towards personal and neighborhood effects, that is, how individuals living along the proposed route would react. Residents and many of their local elected officials did not accept the premise that the objective of the Expressway should be evaluated as an overall regional utility function; that is, it was not acceptable to increase noise, air contaminants, and other environmental consequences in their neighborhoods in order to improve traffic in other neighborhoods. They cared about the impact on them, their families, their friends, and their neighbors, not people who lived 5–15 miles away or people who traveled to the New Jersey shore during the summer months. This Driscoll Expressway EIS is an early example of the reductionist perspective that too often characterizes the position of agencies that badly want their project. With sufficient time, project opponents are able to craft arguments that expose the narrow viewpoint and broaden the issues that they argue should be considered.

## Coordination

My efforts to explore this path were not very productive. People's memories have faded. By speaking informally to several officials, consulting media reports, and examining Governor Byrnes's memoirs, I found that the advocates of this project were quite certain of success, and did not take every opportunity to seek out individuals and groups that were likely opponents in order to reach a compromise. They did apparently, however, reach out to local planning staff in some of the municipalities. Indeed, they were, I think, misled into believing that support from technical staff meant support from local elected officials and the local public.

I cannot fault the Driscoll Expressway EIS for failing to understand the impacts of the short-lived oil shortage in 1973 and how it would impact the views of the new Governor Brendan Byrne. Governor Cahill, the previous governor, had approved the proposal, and the proponents assumed, perhaps naively, that his decision was final. The oil crisis clearly impacted Governor Brendan Byrne's decision (personal conversations and media coverage of his formal remarks, as noted above), but there was no oil shortage when this document was prepared.

I am less able to understand why so little consideration was given to the impact on open space, when the public has been extremely interested in preservation and has funded purchases of large tracts of open space with taxpayer and bond monies. The authors of the EIS indicated acreage that would be lost to the highway, and that there would be impacts on plants and animals, but did not give much credence to it. Thirty years later, the debate over Route 92 in New Jersey brought back the same issue with the same result, that is, the proposal was defeated on environmental grounds.

## Accessibility to other stakeholders

Ultimately, the NEPA process exposed the details of the plan to scrutiny by the public and adamantly opposed environmental advocate groups, and to the skeptical eye of a new governor. The new governor understood the bigger picture of sprawl, the population of cities, energy dependence, and others. Overall, my review of the technical elements of the Driscoll Expressway EIS is that, like many projects of that era, it ignored lessons learned from the civil rights movement and later applied to the environmental justice movement, which is to say that satisfactory biological, chemical, and laboratory science, even with good engineering support, will not necessarily overcome broad ethical/moral, political, and social concerns.

## Fate without an EIS

I cannot be certain. However, I believe that this project would have moved forward and ultimately been built without the EIS requirement and the presence of Governor Byrne. These two, in fact, worked in concert. Governor Byrne did not like the idea of this highway, as well as many other proposed highways. He was disposed to oppose it. The presentation of the EIS went badly for project proponents, which provided him with more than sufficient political ammunition to adamantly oppose the proposal. Had Governor William Cahill still been in office, I suspect that the highway would have been built. Governor Cahill had already agreed to support it, and the EIS, despite strong public outcry against the project along part of the route, and environmental group opposition, was remarkably supportive. Unfortunately, Governor Cahill passed away in 1996, so I have not been able to ask him any hindsight questions.

This EIS was a satisfactory planning document for a different era. The agency failed to adjust to the shifting political power realities that were taking place in New Jersey and in many other states. From a technical perspective, the document is satisfactory in almost all parts; but from a communication perspective, it is painful to read and must have infuriated many readers.

# Hudson–Bergen Light Rail system

## History of the proposal

On December 19, 1999, *The Star-Ledger* (the largest-circulation newspaper in New Jersey and sixteenth in the United States) released a public opinion poll of 804 residents that asked the public to reflect on the past twenty-five years of the twentieth century and identify the state's major failures and successes (Zukin 1999). By a margin of two to one, New Jersey residents identified the failure to stop suburban sprawl and to get commuters from automobiles to mass transit as the two major failures. In contrast, by a margin of more than five to one, the five major successes and achievements were: construction of the performing arts center in Newark, the Meadowlands sports complex, development of the Hudson waterfront, renewal of Hoboken, and efforts to clean up the New Jersey shore. Three of these five successes are directly tied to the Hudson–Bergen Light Rail system.

The Hudson–Bergen Light Rail system was a product of public, business, and government growing frustration with congestion and deterioration in older cities and suburbs. The signal political message was the founding of NJ Transit (NJT) in 1979. This progeny of the New Jersey Department of Transportation began by taking over and managing a variety of bus routes, some of which had been successful while others were floundering. In 1983, NJT began operating all of the commuter rail service in New Jersey, with a few exceptions. NJT is the largest public transit system in the nation and the third largest single provider of bus, rail, and light rail transit. The system's weekday daily ridership is now approaching a million. NJT has almost 2500 buses and 1100 commuter rail trains, including double-deckers (NJ Transit 2007). It operates a monorail to Newark Liberty International Airport, and in other ways is extremely active in providing multi-modal transit options.

In an announcement highlighting "record ridership," NJ Transit (2007) underscored its light rail program. "Growth in rail was up 5.4% over the first quarter of FY07, . . . the Hudson–Bergen Light Rail ridership shot up 18.4% over the same period last year." The Hudson–Bergen has had a long and complex political history, and environmental impact has played a major role.

In 1983, Governor Thomas Kean, who this author would characterize as a moderate Republican with a strong interest in cities, suburbs, and rural areas, reflecting on the issues of sprawl and distressed older cities, issued Executive Order 53, which created a Hudson River Waterfront Development Committee to assess ways of reducing traffic congestion, upgrading infrastructure, and redeveloping the waterfront. Governor Kean, along with Governor Byrne and the governors who followed, recognized the strategic location of the area.

The waterfront stretches approximately 18 miles from Fort Lee in Bergen County to Bayonne at the tip of Hudson County (see Figures 2.2 and 2.4); it includes a residential population of close to half a million in eight municipalities. Directly across from Manhattan West side (hence its nickname, the "Gold Coast"), this area was part of the industrial–port complex of the New York–New Jersey region. But, beginning in the 1950s and accelerating into the 1960s, a good deal of the industrial activity closed and railroads that ran along the Hudson River were underutilized or abandoned, which left thousands of abandoned contaminated brownfield properties.

Governor Kean's committee included senior state, county, and local officials, and regional representatives, as well as community representatives. In 1984, Governor Kean directed the New Jersey Department of Transportation to assess transportation options for the Hudson River waterfront. The Department reported that uncoordinated and unplanned decisions were being made that would hinder future redevelopment. Yet, not surprisingly, analysts stumbled over who should control the transportation rights-of-way and who should pay for the upgrades.

The committee also struggled with a decision about whether the Hudson River transportation projects should not start until there was an agreed-upon plan for all of northeast New Jersey. Governor Kean's administration prepared a plan to link the Hudson waterfront, the New Jersey Turnpike, and the New Jersey Meadowlands – the so-called "circle of mobility" concept. The plan, in essence, integrated a variety of already existing proposals for rail links, road extensions, bus ramps, a monorail or trolley, and a new tunnel to New York City. This multipronged effort detracted, temporarily at least, from the focus on the Hudson River waterfront, and brought some agencies and community groups into conflict with one another because they recognized that funds were limited. Debate over the circle of mobility proposal lasted for well over a decade, and other elements of it, in fact, have been built. But threatened with a loss of federal funds because of an inability to reach consensus, the state chose initially to concentrate on the Hudson–Bergen opportunities.

Some of the difficult negotiations were with Conrail, which owned the right-of-way. After lengthy negotiations, an agreement was reached with Conrail. The state purchased Conrail's right-of-way along the Hudson and provided it with an alternative route for its commercial traffic. In 1989, Governor Kean noted that he recognized the importance of the waterfront to

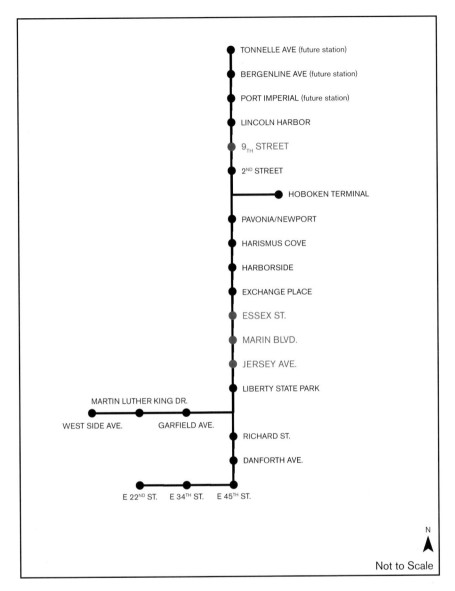

**Figure 2.4**

Light rail stations

the economic development of the area, but that transportation access was essential (Baehr 1989a). Yet part of the argument against the circle of mobility concept was caused by environmental sensitivity toward widening roads and building tunnels through sensitive wetlands areas. Many of those arguments have not been resolved.

## Design concepts

In May 1989, the NJT Board agreed to spend $2 million on a federally required study to assess buses, trolleys, monorails, or automobile options. The draft EIS (NJ Transit 1992) was based on the idea of a thorough review before the investment of federal funds in large public works projects; that is, the analysis was expected to allow local officials, developers, and private citizens to participate (in Hudson and Bergen Counties).

Given the national and state competition for mass transit funding, the no-build alternative and low-cost options were real options, if the interest groups could not agree on a plan. Governor Kean and his advisers recognized that the state did not know with certainty what single option or combination of options would be most effective in garnering support. Indeed, it was not clear that any of the options could improve access for area residents and workers; engender economic development of the Hudson waterfront; preserve wetlands and protect the environment; and lead to a consensus for a transportation plan. The study focused on seven alternative plans for the corridor. These included a no-build scenario that maintained existing transit and roadway services; modest adjustments of transit and traffic (a band-aid approach); adding busways, park-and-ride lots, and a few trolley lines; constructing monorails; building new roadways; and a hybrid of all of the above.

NJ Transit organized a Hudson River Waterfront Advisory Committee to comment via an alternatives analysis/draft EIS (Baehr 1989b). The Advisory Committee was to help NJ Transit identify a "locally preferred alternative (LPA) for which federal and state funding would be sought." After the study began, Martin Robins, project director (see below for interview), noted that he thought light rail might be advantageous, so it was added to the options (Baehr 1991).

This design was to rely on existing right-of-way from Conrail's River Line, and would run from downtown Jersey City west to a park-and-ride lot on Route 440 in Jersey City; next it would go north along the Hudson River waterfront through Hoboken and Weehawken, and under the Palisades through the Weehawken Tunnel to another park-and-ride lot to be located near the junction of Routes 3, 1&9, and 495 (see Figures 2.2 and 2.4 for regional and location diagrams). A little over a month later, Robins unveiled his staff's complete idea, which included a 15-mile light rail system, an express busway from the NJ Turnpike to the Lincoln Tunnel that crosses into New York City, and four large park-and-ride lots with 12,000 parking spaces. A major addition to previous plans was an extension of the light rail north to the Vince Lombardi rest area on the NJ Turnpike in Ridgefield Park. The justification was the opportunity to attract many thousand more additional riders.

Robins expected to finish the plan and hold public hearings by either March or April 1992. Following those hearings, local officials and the NJ Transit Board would agree to a locally preferred alternative and apply to the federal government for a grant to produce a final environmental impact statement (FEIS) and begin final design of the system. Robins hoped to obtain the grant for the FEIS by mid- to late 1992 and begin construction a few years later (Baehr 1991). The FTA approved NJ Transit's draft environmental impact statement (DEIS) in November 1992. The approval came two years after the study was first begun by NJ Transit. However, by that date, the NJ Transit Board had still not adopted the LPA. In February 1993, after 43 months of planning and negotiations, NJ Transit's Board of Directors adopted the LPA for the Hudson River waterfront.

## Conflict and resolution

Interest groups offered dozens of suggestions. For example, a group of north Hudson County municipal officials and a coalition of community, environmental, and transportation advocates criticized the busway proposals as polluting and inefficient, and likely to conflict with light rail (Baehr 1990, 1993). In February 1993, when the NJ Transit Board adopted the LPA for the light rail, a set of four contentious issues remained, that I will summarize.

An extension to Bayonne had strong political support; the issue was cost. Yet the supplemental draft environmental impact statement (SDEIS) found the proposed extension to be much more cost-effective than initially expected, which temporarily solved the problem (see below for interview).

The Jersey City supplemental environmental impact statement (SEIS) highlighted a split in the city between two historical political districts, in which the SEIS became ammunition for both sides. The "City Center" option would have divided the historic Van Vorst neighborhood, whereas the proposed "City South" route would have affected the historic residential Paulus Hook community. Local commercial interests tended to favor the City Center route because it would serve their customers. Transit planners, developers, and many local officials supported the City South path because it would allow access to developable waterfront property. By offering detailed results, the EIS provided the basis of arguments for both positions.

In March 1995, NJ Transit Board member Amy Rosen, who chaired the agency's Engineering and Operations Committee, offered the following comment about the process: "Finding a way through 20 miles of the most densely populated county in the most densely populated state in the nation is not easy," she said. "I am overwhelmed by the leadership of the public officials who had to deal with this" (Baehr 1995). In November 1995, NJ Transit published the SDEIS evaluations for the Bayonne extension and another for the Jersey City alignment. It is fair to say that the details and findings in these documents engendered considerable debate. However, because the EIS process that produced the facts was satisfactory to the parties, they focused their attention on information rather than arguing about a process to create facts. Fewer site-specific noises, dirt, construction-related, and other neighborhood impacts were found to be associated with the City South route, and this route was also less expensive and could be completed more rapidly. These factors were key elements in building a political consensus for the choice of City South.

The third unresolved issue was the construction of a station below Union City in the rail tunnel. This analysis was an environmental assessment because it was not expected to produce notable impacts (see Chapter 7 for an example). The station would be connected to the surface by large elevators. Cost was the key issue (see interview below).

In January 1996, the NJ Transit Board voted 5–0 to accept the supplemental EIS studies for Bayonne, Jersey City, and Union City, thereby amending the original LPA to include the Bayonne extension, Jersey City south alignment, and the Union City station. The amended plans were sent to the Federal Transit Administration (FTA) for review. In August 1996, the FTA published a "Record of Decision" in the *Federal Register*, and gave 30 days for public comment. Assuming no consequential disagreements, the FDA's record of decision meant that it had approved the final EIS.

The fourth unresolved issue was a route through Hoboken. The options were building in the densely developed eastern side of the city, or along a railroad spur on the western side of the city, which was far less developed, indeed clearly in need of redevelopment. What was interesting about this option was that elected officials and community representatives changed their mind as more information became available. Eventually, drawing upon neighborhood environmental activism, the western side of town was chosen (*Jersey Journal* 1997).

Environmental controversies regarding the Bayonne, Jersey City, Union City, and Hoboken project did not end with the EIS and record of decisions. In a densely developed area with centuries of industrial development, excavation and construction found historical settlement, uncovered rodents, and unwanted debris, and produced a considerable number of complaints about construction noise and dust (Torres 1997a,b, 1998a,b, 1999).

Nevertheless, the light rail system opened on April 22, 2000 from Bayonne (34th Street) and Jersey City (Exchange Place) with a spur. Then service was extended to Pavonia–Newport. In 2002, the light rail service continued to Hoboken Terminal, and 2003 it was extended to 22nd Street and Bayonne. It reached Lincoln Harbor In 2004, Port Imperial in 2005, and Union City and North Bergen in 2006 (see Figure 2.4). The current system has forty-eight electrically powered vehicles, each 90 feet long, air-conditioned, with a capacity of sixty-eight seated passengers and standing room for 120 more. Trains operate about every 10 minutes, with lower frequency during off-peak hours and weekends/holidays. Researchers have found that the light rail system has led directly to substantial expansion of residential and commercial development. In each case, the light rail system was linked to bus, ferry, and other rail systems. Daily ridership is about 40,000, which ranks this system below light rail systems in Boston, Los Angeles, Philadelphia, and seven others opened before 2000. Among those opened since 2000, only Houston has more riders.

The Hudson–Bergen line has not reached Bergen County, and original plans suggested this system would have 100,000 riders when completed in 2010 (see Wikipedia: "List of United States light rail systems by ridership," http://en.wikipedia.org/wiki/List_of_United_States_light_rail_systems_by_ridership). Some of this added demand was to come from real estate development, and that has been proceeding (although impacted by the economic slowdown). Other users were to come from finishing the line, which brings us to the last unresolved issue, the extension of the Hudson–Bergen Light Rail system into Bergen County, and how this should be done. The current length is less than 10 miles; if extended to Tenafly, it would be almost 21 miles.

## Scoping Document for the Northern Branch Corridor DEIS

Nearly all the case studies in this book are final or draft EIS documents for large, complicated engineering projects. Here I have made a deliberate choice to examine a scoping document for an EIS. I summarize the twenty-two-page scoping document not because it speaks to environmental protection, but rather because it underscores that the EIS process has become a written exercise in diplomacy in many cases where the proposed project is in a densely developed area (US Department of Transportation 2007). Unlike the Driscoll Expressway EIS, which frankly dismisses public concerns as misplaced and even harmful to the environment and public, this scoping document reads like a diplomatic exercise. It begins by succinctly describing the project and its historical context. The second page is a legible map of the study area. Then the purpose of the scoping document is described. Notable are the last two sentences of the first paragraph:

> The purpose of the scoping document is to provide information to the public and agencies regarding the Northern Branch DEIS process, issues, alternatives and methodologies. The broader purpose of the scoping process is to provide an opportunity for the public and agencies to comment on and provide input to the Northern branch DEIS as it is initiated.
>
> (*ibid.*, p. 3)

In contrast, as noted above, the Driscoll Expressway EIS pays virtually no attention to the public: the first sentence of the cover letter from the consultants sets the tone with respect to the public: "In May 1972 the citizens of New Jersey, by formal action of their elected representatives, instructed the New Jersey Turnpike Authority to proceed with the planning, design and construction of a limited access highway through Ocean, Monmouth and Middlesex Counties" (New Jersey Turnpike Authority 1972, page not numbered). The last line in the penultimate paragraph of the transmittal letter leaves very little space for public objections:

> Based on this evaluation, we have not identified any significant damage that might occur to the environment as a result of the proposed Expressway. Further, appropriate safeguards and controls can be included in the design and implemented during construction so that the Governor Alfred E. Driscoll Expressway will actually enhance and improve many areas through which it traverses.
>
> (New Jersey Turnpike Authority 1972,
> transmittal letter)

Rather than providing no alternatives, the options presented in the Northern Branch Corridor DEIS are labeled as "preliminary alternatives." On page 6, the document uses the words "rethink," "new perspective," and "rethought" to discuss the process.

Readers of this book realize that assertions of "public" input need not correspond with decisions. But in this case, appearance seems to be equal to intention, which means that elected officials and agency representatives realize that this project has very little chance of success without a strong public mandate. The scoping document invites the public to consult the project website and post questions, register for testimony, and sign up for the website mailing list. Contact people are provided by name, with phone and fax numbers.

Six goals and objectives are stated. The first goal is to "meet the needs of travelers in the project area" (US Department of Transportation 2007, p. 13). The specific objectives for this goal are "attract riders to transit, improved travel time, improved convenience, provide more options for travelers, and improve services for low income/minority/transit dependent travelers." This list of objectives is remarkably attuned to the complaints of the New Jersey public noted in the 1999 survey discussed above, and in following surveys, and is consistent with the concept of new urbanism and environmental justice.

The remaining five goals include advanced cost-effective transit solutions; encourage economic growth; improve regional access; reduce roadway congestion; and enhance the transit network. While the labels are not obviously directed precisely at securing public support, the tone of the language is clearly an attempt to be candid, an attitude that risk-communication studies show that the public appreciates. For example, the following is stated about advancing cost-effective transit solutions:

> The Northern Branch project should allow for future transit expansion while at the same time provides a solution that is affordable to construct. With limited capital funds, the ability to advance projects in phases helps

to keep the project affordable. Project scalability allows projects to be constructed without precluding future expansion projects. One of the criteria
on which the Northern Branch project will be evaluated is the degree to
which one phase of the project integrates into a more global planning effort
for transportation improvement in the region.

(US Department of Transportation 2007, p. 14)

With regard to roadway congestion, there are several statements that
neither promise the public a perfect solution, nor abandon them.

> Major regional highways in the project area are heavily congested. There
> are a limited number of major highways, each serving intra-county and
> regional travel needs. Congestion in Bergen County is a growing problem,
> is likely to become more serious in the future. Transit strategies are unlikely
> to substantially reduce congestion, but can provide useful new travel alter-
> natives for travelers trying to avoid congestion.

(ibid., p. 14)

The project alternatives are not discussed until page 16 of this twenty-one-
page document. They include a no-build alternative, diesel–multiple-vehicle
alternatives, and electric light rail vehicle alternatives. While there had been
considerable discussion of these alternatives, there is no evidence that the EIS
had a predetermined outcome. Table 2.1 lists the general and more specific
impacts, which look remarkably like those for almost every other urban transportation project I have seen.

Following the same theme, the reader will not be surprised by the content of
the last two pages. They describe the "public involvement program." Included
are newsletters, a study website, a citizen's liaison committee, breakout sessions,
agency coordination, small town meetings, and scoping meetings/public hearings. In fact, many of the communities have begun their own evaluations of this
extension (Borough of Tenafly 2009).

## Conversations with Martin Robins

Martin Robins was a colleague for many years, and I know of no-one who
better understands the reality of creating, building, and managing a rail system
in a complex political environment. His career has involved numerous challenging efforts. Martin Robins's career in transportation planning and policy
extends more than three decades. He has the capacity to cut through details
and get to the key issues.

An attorney, Robins was Director of the Alan M. Voorhees Transportation
Center at Rutgers University, and now advises several Rutgers transportation
research centers. Prior to that, he was project director of Access to the Region's

**Table 2.1** EIS elements in the Hudson–Bergen Light Rail study

Impact category

| Transportation | Natural environment | Built environment | Environmental justice | Construction | Indirect and cumulative |
|---|---|---|---|---|---|
| • Air quality<br>• Noise and vibration<br>• Traffic, parking, transit, pedestrians, freight rail<br>• Electric and magnetic fields<br>• Safety and security | • Water quality<br>• Wetlands<br>• Flooding<br>• Navigable waterways and coastal zones<br>• Ecologically sensitive areas<br>• Endangered species<br>• Hazardous-waste | • Land acquisition and displacements<br>• Land use, zoning and economic development<br>• Consistency with local plans<br>• Historic properties and resources<br>• Parkland<br>• Archaeology<br>• Aesthetics<br>• Community disruption | • Community disruption for minority and low-income communities<br>• Access for minority and low-income communities to proposed project services | • Storage of construction materials<br>• Noise impacts during construction<br>• Traffic, parking, transit, pedestrian, and freight rail impacts during construction | • Impacts from other regional transportation on land-use developments |

Core, a three-agency partnership focused on the need for a new rail transit tunnel between midtown Manhattan and northern New Jersey. Robins had been director of the Port Authority of New York and New Jersey Planning and Development Department, as well as deputy executive director of NJT. Most important from the perspective of this EIS, he was director of NJT Waterfront Transportation Office, which planned the Hudson–Bergen Light Rail line.

Robins had no difficulty recalling the issues, facts, scientific debates, and politics of the Hudson–Bergen Light Rail project. I spoke to him about this project on various occasions, and we had a longer meeting on April 20, 2009. He agreed with my belief that local elected officials, their staff, and community groups pressed experts to make choices that were, in the long run, better for the environment and the public. When Robins began working on options for the Hudson–Bergen corridor, light rail was one possibility and bus lanes was another. The NJ Transit unit he headed met about a half dozen times with a citizens' advisory committee. Public sentiment was for light rail and strongly against the noise, dirt, and smells of more cars and buses.

With regard to light rail routes, he has explained on several occasions that local officials strongly influenced the routes. For example, Bayonne's mayor and elected state representatives pressed for a location along the less densely inhabited eastern border of the long, narrow city of approximately 70,000 people. Robins's planning staff advocated a route through the populous, developed core of the city in order to generate more ridership and reduce automobile traffic. Local political pressure led to the selection of the eastern route, which, in the long run, he acknowledges turned out to be a better choice because the US military closed the vast Military Ocean Terminal facility in Bayonne. The location of the light rail has helped spur redevelopment of that spacious area for housing, commercial activities, and tourism. The location preferred by his staff would have provided less of a spark.

Robins and this author both enjoyed recounting a selection of the Jersey City route. Western and Eastern routes were proposed. It is not unkind to label the Western route as passing through an area that was a mess. The population was impoverished and redevelopment was hindered by massive hexavalent chromium contamination, a residual of deposition from PPG Industries, Occidental Petroleum, and Honeywell International Corporation. Honeywell has spent an estimated $400 million to remediate the area. Robins reiterated that he wanted to extend west from the route that was ultimately chosen, over Route 440, Roosevelt Stadium, and Society Hill in Jersey City. But the line was not extended because the course of the chromium clean-up had not been legally resolved.

The West Side alignment in Jersey City presented an interesting case of citizen preference affecting the profile of the route. The early design for the West Side alignment would have placed the light rail in an abandoned railroad cut 20 feet below the surface. The activist local population, mainly female, beset by poverty, underinvestment, and crime, argued that a sub-surface design would engender criminal activity; their arguments were accepted by the

Robins team and the line's profile was lifted to street level at the site of their key neighborhood station.

The route in downtown Jersey City was strongly influenced by the development arm of Colgate-Palmolive and middle-class residents of the Van Vorst neighborhood. The light rail line now passes directly through property owned by Colgate, which strongly influenced city officials to locate the rail line on their property and avoided the hostile Van Vorst brownstone neighborhood. The Robins team had favored a different route, which served an old merchant street, passed by the edge of the Van Vorst neighborhood, and connected well with the Port Authority Trans-Hudson (PATH) Grove Street station, served by trains destined for both midtown and lower Manhattan. Nevertheless, the routing selected through the Colgate property and other vacant land has attracted a prodigious amount of residential and office development.

The alignment in Hoboken was controversial, with the Robins Team favoring an easterly alignment along the waterfront, serving newly emerging developments. Citizen groups favored a westerly alignment along a railroad right-of-way that was surrounded by junk yards and the Palisades cliffs. The resulting westerly alignment, the choice of which was influenced by environmentalists and political considerations, has attracted far more economic development than had been anticipated. Hoboken has been the city in this area with the strongest environmental protection advocates. The environmental community opposed the easterly light rail alignment because it would cut somewhat into a historic rock formation (Castle Rock) and would have occupied the occasional recreational resource of Frank Sinatra Drive. They wanted light rail, opposed any more bus lanes, and called for a western route through Hoboken. The choice they argued for, in fact, has worked well, as this formerly dormant west side of Hoboken has sprung to life and the area has been developing rapidly.

The construction of the West New York (New Jersey) station was a challenge for technically oriented planners. State legislator Robert Menendez (now US Senator from New Jersey) wanted a station in the Palisades tunnel under Union City and West New York. This meant drilling vertically 150 feet from the densely populated street atop the Palisades through hard rock to a station in the tunnel. The cost was high, and the experts initially were concerned about the station elevators' capacity to process users quickly enough. Once pressed by Menendez, the engineers found that the station was feasible, and this Bergenline Avenue station has been one of the most successful in the entire line. Large numbers of people use it for work, and the local population uses the station on weekends to visit otherwise inaccessible shopping areas along the light rail line. The estimated actual use has more than doubled initial ridership estimates.

Robins, an attorney, planner, and transportation expert, sees the EIS process in this case as a regional planning mechanism. Without it, the project very likely would have failed. The EIS required gathering and evaluating a diverse set of data, and those data were used to craft a design that fits local needs.

An excellent communicator and writer, Robins worked hard to make the document readable, specifically to deliver a clear story in the executive summary and throughout the text. He recalled spending hours editing and re-editing until the message was clear.

Reacting to my question about "cumulative effects," Robins indicated that his team spent a great deal of time considering these, especially when the community would propose a change in the design. While they emphasized each segment of the line, the entire project had to offer a clear net benefit in terms of reducing air pollution and noise from cars, increasing ridership, and economic development in the region as a whole.

By now, the reader may conclude that this evaluation seems a little too good to be true; that the author, supported by Martin Robins, must surely have exaggerated the political and public input that influenced this light rail system. Hence we also discussed the generalizability of the Hudson part of the Hudson–Bergen line to other locations. Hudson County has had a history of powerful political leaders that cannot be ignored by federal agencies. This history means that citizen groups and elected officials can make a persuasive case, and federal and state officials would be foolish to ignore them. The step-by-step planning process required by the EIS fitted the political environment of this area like a glove. The Bergen part of the project has not moved ahead, and Bergen County elected officials are perhaps less powerfully placed and less reliant on their political leadership than their Hudson counterparts. Is this the reason why that part of the project has not been completed? I doubt if it is the major one – but it is a reason.

A second important point is that, compared with many other projects with which the author is familiar, water and air pollution and solid waste management were not such powerful factors in this case as they often are. Social and economic impacts were consistently important, whereas traditional environmental factors were of greatest importance only in Jersey City and Hoboken.

# Evaluation of the five questions

## Information

There have been more than two dozen EIS documents, primarily because of the proposed extensions of the light rail system. I have not read every page of every one of these documents. Some of the information, such as concern about chromium contamination, is provided in substantial detail, supported by a good deal of scientific information. Other information is treated more lightly. My sense is that the agency focused much of its attention on the issues that concerned local elected officials and community groups. In other words, the scoping process helped define the focus of the environmental assessments. Clearly, the staff attempted to explain the information without unduly steering it, although it is hard not to be directive in some of these cases.

## Comprehensiveness

The document includes a broad spectrum of environmental, economic, and social considerations, perhaps more than I have ever seen in a single project. The emphasis is on land use and transportation planning. The documents consider cumulative effects, especially on land use and economic redevelopment. If there is any shortcoming, it is perhaps that environmental issues were less developed than were economic and social ones, which concurs with Martin Robins's statements. However, that reality may relate directly to the on-the-ground reality of the area. A great deal has been expected of the approximately $2 billion investment in mass transit. Proponents created some ambitious goals (over 100,000 daily commuters, 33 million square feet of commercial development, 58,000 new jobs, 40,000 new housing units, and cleaner air). It is too early to measure the full impacts. However, initial evidence suggests that, notwithstanding the international economic slide, the Hudson–Bergen project has been successful. For example, Goldman Sachs has built a forty-two-story tower on the former Colgate-Palmolive site, and thousands of new rental sites and some for-sale housing has been constructed within half a mile of the Jersey City rail stations. A major hospital in Jersey City relocated to be along the line. Surveys show that residents living within half a mile of the stations are far less dependent on automobiles than their counterparts elsewhere (Wells and Robins 2006).

## Coordination

There was a major effort to provide information, formally and informally, to other federal, state, and local agencies. I have never seen an EIS process that had more meetings, documents, and opportunities for access to the process, albeit elected officials' access was the most important.

## Accessibility to other stakeholders

Looking back at the building of the Hudson–Bergen Light Rail system, I would not deny the usual political influence in the process. Key elected officials, such as Governors Byrne, Kean, Florio, and Whitman, local members of the US Congress such as Congressmen Roe and Menendez, as well as the Commissioners of the New Jersey Department of Transportation, had the power – which they used – to influence this set of projects. They could and did exercise vetoes and strongly influence options. If you ride the Hudson–Bergen line, you will find some unusual changes of direction that would not have emerged without political influence. Yet, looking back at decades of scoping, draft, final, and supplementary EISs regarding the Hudson–Bergen Light Rail system, the author concludes that the EIS process helped New Jersey officials form a consensus that otherwise might not have materialized in a densely developed and highly politicized environment. The EIS process constituted the evidence to undermine many of the projects that were part of the initial highway elements of the redevelopment plan for the Gold Coast. The process also allowed technical experts to prepare the environmental impact estimates, which were critiqued by proponents of one or another option. The EISs served as fodder for the political process. The EIS process forced local interest groups to bring their concerns to the table in a timely and controlled fashion. It allowed decision-makers to stand back and watch the survival of the politically fittest ideas emerge. I think that less credible scientific information would have been used were it not for the EIS process; and, despite the charge that the EIS has been abused by interest groups to obstruct projects, in this case it is hard to believe that these projects would have been built with relatively strong public and political support without the EIS process.

## Fate without an EIS

I believe this project would have not have moved forward without the EIS requirement. The process forced to the table people who otherwise might have sabotaged the project behind the scenes. It was molded to fit local needs, and it illustrates the idea of a federal government planning process (see Chapter 8).

# 3 Ellis Island, New York Harbor: time closes in on a national cultural treasure

## Introduction

More than a dozen national parks are located in New York Harbor and immediately adjacent areas of New York City and northern New Jersey. These include the Statue of Liberty, its northern neighbor Ellis Island about half a mile away, its southeastern neighbor Governors Island, the Gateway National Recreation area in Jamaica Bay, and the African Burial Ground National Monument in lower Manhattan (Figure 3.1). Ellis Island, the focus of this chapter, is one of the most important cultural sites in the United States, and arguably is the one most in need of attention because of physical deterioration of the physical structures on the South Island. This chapter examines the final EIS documents and management plans for Ellis Island as an illustration of an EIS process that has focused on cultural and historical attributes.

From January 1, 1892 to November 12, 1954, Ellis Island was the major port of entry for migrants. Over 12 million people entered the United States through Ellis Island, the vast majority disembarking from cramped quarters in steamships. Twelve million may not seem like a large number in a population of over 300 million US residents in 2009. Yet about 40% of Americans, including this author, can trace their family history through Ellis Island.

When the Island's immigration function ended, buildings on the 27.5-acre site began to degrade. In 1965, President Lyndon Johnson added Ellis Island

New Jersey
Liberty State Park
Hudson River
Manhattan
Ellis Island National Monument
Liberty Island
Governors Island
Brooklyn
N
Imagery Source: Bing Maps; ArcGIS Online: http://www.arcgis.com
Copyright:(c) 2010 Microsoft Corporation and its data suppliers
0 750 1,500 3,000 Feet

**Figure 3.1**

Ellis Island and environs

to the national park system as a part of the Statue of Liberty National Monument. With federal funding, the main building on the north side (see Figure 3.2) was reopened to the public in 1990 as the Ellis Island Immigration Museum, and then two adjacent buildings were rehabilitated as offices. The National Park Service (NPS) maintains and has stabilized the south side, and one of the buildings has been restored. But the remaining twenty-nine buildings, which include the medical facilities for immigrants, probably will need to be demolished in 10–20 years unless there is an intervention (see Figures 3.3 and 3.4 for illustrations of deterioration).

The question before the US government and its elected officials is: Is immigration a sufficiently powerful force in the United States to warrant rehabilitation and reuse of the remainder of Ellis Island in ways that will honor its historic role? The number of immigrants in the United States has been rising, reaching record numbers in some cities and states. Many of these immigrant groups are minorities with relatively high fertility rates. Some suggest that a "browning of America" is occurring and that eventually non-Hispanic Whites will be the ethnic minority (*Time Magazine* 1990). Padilla (1977) observed that if these trends continue, it is possible that, by the year 2040, one in four US residents will be foreign-born. The US Census Bureau projected that in 2050 only half of the US population will be non-Hispanic Whites (Malone *et al.* 2003). In August 2008, it changed the estimate to 2042. In other words, the process is moving more quickly than had been anticipated.

New Jersey and New York border Ellis Island, and rank second and third, respectively, in their proportion of foreign-born residents (California ranks first). New Jersey's and New York's foreign-born populations are the most diverse in terms of the number of ethnic groups (Lapham 1993). These states are a forerunner of the changing demographics of the United States.

But assuming that immigrants should be honored by expanding the Ellis Island shrine is an arguable contention. Controversy surrounds the impact of immigrants on the economy, and especially jobs (Stryker 1987; Schneider 2007). Do immigrants take jobs away from Americans born in the country? Or do they work in jobs that other Americans will not? Given that many immigrants are highly successful, as measured by family income and education, do immigrants, on balance, create jobs for all Americans? Public perception of whether or not immigration is "good" or "bad" for the nation clusters in certain locations and within certain groups, and has become a political issue in some locations.

While some criticize immigrants, the image of the United States as a melting pot is rooted in the continual influx of immigrants, their sacrifice and hard work, and eventual assimilation of many into society (Glazer and Moynihan 1970). Indeed, some scholars credit high levels of civic participation among some US populations to immigrants and to intermarriage of immigrants (Hirschman 2005). The *New York Daily News* holds a Fourth of July essay contest. The winners in 2009 were five 8- to 12-year-olds, and their essays

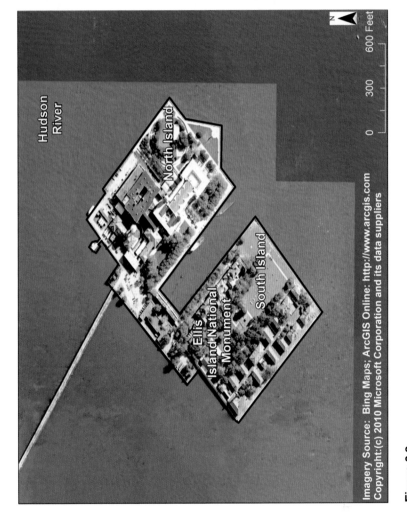

Imagery Source: Bing Maps; ArcGIS Online: http://www.arcgis.com
Copyright:(c) 2010 Microsoft Corporation and its data suppliers

**Figure 3.2**

Ellis Island

**Figure 3.3**

Deteriorating structures on South Island

**Figure 3.4**

Deteriorating structures on South Island

spoke with reverence about their grandparents or parents who migrated from Italy, the Philippines, and Poland. A 12-year-old summarized their views: "As an immigrant, Ellis Island symbolizes the main gate of entry of liberty and hope" (*New York Daily News* 2009).

Yet public perception of immigrants is mixed (Allport 1954; Buck *et al.* 2003). Support for immigrants is positively associated with the strength of the US economy, with greater support in times of economic growth and low unemployment (Espenshade and Hempstead 1996; Citrin *et al.* 1997; Espenshade 1997; Burns and Gimpel 2000; Haubert and Fussell 2006). On the individual level, those with less income and educational attainment are less likely to favor immigration (Chandler and Yung-mei 2001; Scheve and Slaughter 2001). Historical trends suggest that the public has become increasingly negative towards immigration, with a clear concern about levels of illegal immigration (Lapinski *et al.* 1997; Pantoja 2006; Schneider 2007).

Blumer posed a theory of symbolic interactionism, which states that social interactions are crucial in forming opinions of others (Stryker 1987). A recent meta-analysis of research using group contact theory to explain prejudice found that 93% of the studies concluded that contact lessens opposition to and prejudice against immigrants (Pettigrew and Tropp 2005). Yet some assert that intergroup contact increases animosity towards minority groups. For example, over 60 years ago, Key (1949) concluded that the concentration of African Americans at the community level was strongly related to Whites' negative perceptions of that group. Similarly, other quantitative studies have echoed these findings (for a review see Forbes 1997).

There are reasons for a lack of consensus about the impact of intergroup contact. Perhaps there is a threshold beyond which interactions tip toward more understanding, and another threshold that tips toward more hostility. Some kinds of interaction likely are more critical than others. Hewstone *et al.* (2002) reported that a variety of factors may mitigate or influence contacts, including group size, perceptions of threats, and personality differences. Amin (2002) cautions against making simplistic assumptions and highlights the importance of factors such as social exclusion and ethnic isolation, insensitive policing, and institutional ignorance. Overall, there is a general consensus in the literature that interactions among different races and nationalities increase acceptance, but that consensus is not universally accepted.

In this economic and political environment, does it make sense to highlight the history of immigration that Ellis Island represents? Ellis Island is the rarest of public places, perhaps the most significant shrine to the promise and heartbreak of immigration in the United States. Ellis Island was the place that would turn back some 2% of those who made the voyage. Others became sick and died on Ellis Island. For this less fortunate group, Ellis Island was the "Island of Tears." For the vast majority, Ellis Island was the place where they disembarked for resettlement and integration into society. Perhaps the United States' response to the rehabilitation of Ellis Island, especially during a period when budgets are stressed, is a symbolic message about how the United States views immigration. The issue of immigration in today's context is the major reason I chose Ellis Island for a case study of cultural and historical artifacts.

A second reason for choosing this case study is local politics. New York and New Jersey were locked in a dispute about sovereignty over Ellis Island for over 150 years. The two states drew a line in the Hudson River, which gave New York jurisdiction over the original 3 acres that constituted Ellis Island. But more space was needed to receive additional immigrants, and fill was used to increase the island's size to 27.5 acres. With the island as a source of prestige and potential tourism revenue, the two states and their elected officials have exchanged verbal barbs. For example, during the early 1990s, Senator Frank Lautenberg of New Jersey had set aside funds to build a permanent bridge from New Jersey to Ellis Island (MacFarquhar 1995) located a quarter-mile away. His staff had estimated that 115,000 more people could visit the site if a permanent bridge existed. Nita Lowrey, a New York Congresswoman, contended that a permanent bridge would "seriously mar a great monument." She called for all visits to occur by boat, the way the immigrants did. New Jersey officials responded that the Circle Line Company, stationed in New York City, was behind New York's opposition to the bridge. New York officials criticized New Jersey as a greedy latecomer that wanted the revenue from Ellis Island. New Jersey officials produced evidence that the President of Circle Line donated money to Senator Lautenberg's opponent in the senatorial election.

In 1998, the US Supreme Court settled the geographical turf issue by ruling, in a six-to-three decision, that the 1834 compact between the two states gave New York sovereignty over the original 3 acres, but that the remaining 24+ acres were granted to New Jersey (Shaw 1998). The EIS allows us to see if the New Jersey–New York tension has subsided in light of the threat to the South Island.

The third reason for focusing on Ellis Island is the degradation of the south side. The main building on the north side was restored and opened as a museum in 1990. The challenge of the EIS as a planning tool was to develop alternatives for the thirty buildings that were used for hospitals, quarantine, and similar activities. The planners must try to satisfy the surrounding states and cities as well as pro- and anti immigration groups, and their plans have to be affordable.

# Ellis Island Development Program: the plan and EIS

## History of the proposal

After the immigration function was discontinued in 1954, thirty vacant buildings (approximately 375,000 square feet) began to deteriorate. Federal, state, and private stakeholders pressed the NPS to at least temporarily stabilize the buildings and develop a plan that would rehabilitate the structures, expand tourist experiences, and provide sufficient economic resources to maintain the structures. The NPS responded with a short-term rehabilitation and stabilization program and a long-range plan captured in the Development Concept Plan (National Park Service 2005). The document drew heavily on earlier work that includes both the Statue of Liberty and Ellis Island. The NPS's general objectives were to protect Liberty Island's and Ellis Island's cultural, historical, and natural resources. Protecting these resources are the generic objectives of many cultural–historical rehabilitation site projects. The more detailed site-specific objectives for Ellis Island were to rehabilitate and reuse the island's award-winning Beaux Arts campus of integrated brick, stucco, and tile structures with connecting corridors of masonry and glass. As part of that effort, the NPS was to design a landscape of mature trees, grass, and other landscaping that would complement Ellis Island's historic themes. These structural and landscape designs were to reflect the island's special history and yet relate directly to contemporary migration, public health, and education. This combination of themes hopefully would garner public interest beyond the current museum and thereby attract financial and political support to limit reliance on governmental revenues.

To complicate this challenging assignment, in light of the terrorist attacks of 2001 at the World Trade Center a little over a mile away from Ellis Island, the NPS must control access to the island as a potential target and provide security for visitors, staff, and resources of the island. The NPS has to respond immediately to emergencies, which, in turn, required reconsideration of access and egress options.

In essence, the 2005 EIS analyzes three options.

## No-action alternative

Option 1, the no-action alternative, would stabilize the vacant buildings and the temporary bridge to New Jersey. In essence, the no-action plan is a band-aid solution that the NPS has been implementing. The NPS would fully stabilize abandoned and unused buildings on the south side. In fact, much of this work had been completed when the EIS was published. The NPS uses temporary ventilated wood and plexiglass window panels to reduce water infiltration and increase air flow through the buildings. The original clay tiles are removed and temporarily replaced by asphalt shingle roofs. Stone and brick masonry is repaired, as are failing exterior or interior walls and weight-bearing structures. The corridors are cleared and repaired and the utilities are fixed to support minimal use. Gutters, leaders, and other repairs are made to control rain and snow melt. Vegetation damaging the structures is removed, and inside the buildings, debris and hazardous materials are hauled away.

This package of band-aid solutions, costing more than $50 million, is buying 10–15 years for a sustainable plan to be developed and implemented before the building deterioration becomes irreversible, requiring demolition.

The general public would not be allowed in these no-action alternative stabilized structures. The no-action alternative solution requires maintenance of a security program, including the existing temporary construction bridge to New Jersey. When the temporary bridge is no longer functional, access will have to be by ferry or barge. In short, in 10–20 years the no-action option means the complete or almost complete loss of these thirty buildings. Should this occur, presumably the NPS or another organization at least will take photos to document the nations' immigration history on these two islands.

Before summarizing alternatives 2 (day-only option) and 3 (day/night option), to avoid repetition, I summarize some elements common to both alternatives. Both preserve key south-side Ellis Island buildings and their surrounding environment, and provide options for economically sustainable adaptive reuse proposals. Both continue to rely on ferry boats by both day and overnight visitors. Given that the ferry trips are more expensive than a bridge from New Jersey, the EIS calls for subsidized ferry fares for low-income visitors and schoolchildren, reduced-fare days, and special passes. A third common element is construction of a permanent bridge from New Jersey that would be used for emergency response and evacuation; and for construction, main-tenance, deliveries, and other operations. In the case of the preferred option (day/night), the bridge could be used to pick up and drop off conference center guests. Another commonality is infrastructure improvements to support the use of the thirty buildings and surrounding lands. Also, both alternatives require screening of visitors, packages, and vehicles, as well as capacity for emergency evacuations. The two alternatives also preclude several actions: deliberate demolition of any structures; construction of major new structures; reuse of buildings for dominant retail and/or commercial purposes; and pedestrian use of the service bridge. Lastly, both alternatives assume increases in the NPS budget for the island and from not-for-profit partners.

## Preferred alternative: day and night use

The EIS preferred alternative is one that the NPS believes is most likely to be economically sustainable. The key element is an "Ellis Island Institute" that would be developed and managed by a nonprofit partner. The Institute would offer cultural and educational programs and activities. It would also include a policy research center.

Notably, a conference facility and overnight accommodations would be developed, financed, and managed by a professional hospitality business part-ner working with the nonprofit partner. The facility would host meetings, retreats, and workshops focusing on US immigration, world migration, public health, cultural and ethnic diversity, and family history.

A signature add-on of the preferred day/night alternative is the increased use of the site for overnight lodging and related facilities for conference parti-cipants. This option includes conferencing, internet and other communications, and other technological systems essential to make a such a facility competitive. The NPS has been working with Save Ellis Island, Inc. (SEI) to have this organization be the nonprofit partner for the implementation of the preferred alternative. Over $6 million was raised and used to restore the Ferry Building on the south island, which was reopened in April 2007. This preferred option would rehabilitate and reuse the twenty-nine other buildings over a 5- to 7-year period through a combination of government appropriations, private financing, and philanthropy. The preferred option rests on the assumption that a high-level facility focusing on migration, public health, and related topics located minutes away from lower Manhattan and northern New Jersey, and with a truly remarkable view of the entire New York harbor area, would attract visitors for day trips and conferences.

## Second alternative: day use

This option does not provide for overnight accommodations or commercial office space. Yet office space for nonprofit organizations is part of this design. The plan calls for restoration of Ellis Island's historic buildings and landscape over a longer, 10–15-year period, using private fundraising efforts and federal appropriations. The NPS proposes a combination of partnerships, cooperators, and traditional concession operations to provide visitor services, programs, and maintenance of buildings. Building exteriors would be restored and interiors completed to "core and shell" condition. Tenants would provide interior finishes. NPS calls for preservation of one or more selected interior spaces left in "ruin-like" condition for future research and interpretation. Outdoor areas would be tied to the island's history. Visitor services would be provided through concession agreements.

## EIS Elements

Table 3.1 (p. 64) lists the impact topics addressed in the final EIS.

The EIS did not consider the following impact topics because no impacts from the actions were expected: wetlands (none exist at the park); the 35,000-item museum collections; and environmental justice. The last of these three exclusions seems odd in light of the fact that access to the site via a bridge versus a ferry has been debated, and part of that debate is about poor people; the vast majority in this region are African American and Latino. Indeed, the topic is not ignored, but rather covered under access to Ellis Island.

**Table 3.1** EIS elements in the Ellis Island study

| Transportation | Natural environment | Built environment | Indirect and cumulative |
|---|---|---|---|
| Access to ferry terminals | Marine sediments | Historic architectural resources | Hazardous materials |
| Circulation | Geologic resources and soil | Archaeological resources | Noise |
| Access to Ellis Island | Floodplains | Cultural landscape | Visitor experience |
| Parking | Fish | | Tourism |
| | Vegetation/ threatened and endangered plants | | Administration |
| | Wildlife/threatened and endangered wildlife | | Ellis Island infrastructure |
| | Surface water | | |
| | Groundwater | | |
| | Air quality | | |

## Preservation of historical architectural, cultural, and associated marine resources

The no-action alternative clearly is different from the other two alternatives because it does not provide a plan to preserve Ellis Island's historical treasures. Many of the buildings on the National Register of Historic Places (National Register) would not be salvageable in 10–20 years. Options 2 and 3 preserve the historic properties.

In addition, the no-action alternative eventually results in the loss of the bridge connecting Ellis Island with New Jersey. Should there be a fire, older historic properties on Ellis Island surely would be at higher risk of fire damage before fire-fighting equipment could reach the island, unless a boat was used for fire fighting. The bridge proposed under the alternatives arguably limits the water use of the small water body between the island and the mainland (about a quarter mile).

The two preferred alternatives require digging into the terrestrial and marine environments, and could discover buried artifacts. Under the worst-case scenario, some of these would be destroyed or damaged.

## Natural resources

Natural resource impacts are minor. The three alternatives require removing the bridge and closing the gap in the existing floodwall linking Liberty State Park to New Jersey. Marine sediments would be disturbed by removing the pilings in the current bridge, and the two preferred alternatives would disturb sediments when new pilings were added for a permanent bridge. The existing bridge – indeed, any new bridge under the two alternatives – could be flooded, and a bridge slightly increases the likelihood of flooding on the island.

On-site construction and adding a permanent bridge and access roads would disturb vegetation and add impervious cover. The island has several state-protected plant species that the analysts argue can be protected by avoiding them and replanting, if required. With regard to fish, removal of the bridge and building a new bridge would temporarily impede fish navigation, increase turbidity, and probably resuspend toxins now residing at the bottom of the channel. The construction would also temporarily impact wildlife, including a bird species protected by the State of New Jersey.

During construction, emissions and noise from construction equipment would increase, for longer under the two alternatives than for the no-action plan. If either of alternatives 2 or 3 is built, the analysts expect air emissions to increase 5%, a negligible change. Noise impacts are predicted to follow the same patterns, in other words, increases during construction and little impact thereafter. Overall, environmental resource impacts do not appear to be a major issue.

## Social and economic resources

The expectation is that, under the no-action alternative, tourism at Ellis Island would increase slightly. Alternatives 2 and 3 are predicted to result in a benefit to tourism from more visitors to and around Ellis Island, as well as increased demand for lodging in northern New Jersey and Manhattan.

The report estimates that removing the current bridge could increase emergency response times up to tenfold, which could be a serious problem for someone requiring emergency services. A permanent bridge also would help site workers to cross from New Jersey. Without a bridge, ferry traffic would have to compensate, leading to more automobile traffic and parking at the ferry sites. Liberty State Park, where a permanent bridge would connect to the island, would experience more traffic and parking.

The analysts assert that the day/night alternative would substantially increase social benefits by increasing visitor access to more of the island's historical legacy, as well as the expansion of interpretive programming. The proposed conference facility with overnight lodging likely would result in a major benefit to some visitor experiences at Ellis Island. With the exception of overnight lodging accommodations, similar benefits to the visitor experience are predicted for the day-only option. In the short-run, visitors would notice more noise and traffic at the rehabilitated areas. Lastly, either of options 2 or 3 would require upgrading the utilities and infrastructure.

Overall, more than a half century after the immigration function ended, key parts of the Ellis Island legacy are essentially awaiting their fate. The NPS argues that "a window of opportunity exists – interim stabilization measures, combined with the determination of highly motivated nonprofit and governmental partners, have created the opportunity for what may be the last best chance to save these historic treasures" (National Park Service 2005, p. xiii). The authors of the EIS assert that their "plan honors the legacy of Ellis Island and sets the stage for the restoration and adaptive reuse of the entire historic immigration station complex" (*ibid.*, p. xiii). The NPS considered reuse options that would stop deterioration of the thirty buildings and allow activities that are economically sustainable, and would not be perceived as turning over this unique part of US history to business interests for exploitation.

Doubtless, restoring the thirty buildings and landscaping would provide visitors with a snapshot of what this tiny city within its largest city did to welcome, screen, and care for millions of immigrants. The plan allows the reader to envision a place that would be more than another museum; stepping on Ellis Island immediately causes an emotional reaction for many Americans, creating a credible place to explore immigration in all its facets – racial and ethnic tolerance, civic responsibility, and provision of public health.

I have spent decades on projects to manage and destroy bombs and their wastes (Chapters 5 and 6). Hence my first reaction to the tables estimating the costs required to reuse Ellis Island was to discount the cost issue as trivial. I assumed the EIS controversy, if any, would be about how to reuse the site. But, after I read the EIS, I left persuaded that so much coalition building had already been done on the value of the legacy that dollars were the key issue. The $100 million plus are modest compared with government investments in support of every other project studied in this book. Yet the NPS is not the Department of Defense or Department of Energy, nor even the Department of Transportation, so the costs involved are high for the responsible government parties.

I characterize the presentation of costs of the three alternatives in this document as underwhelming. If the intent of the authors was only to provide general descriptions of the two options and the costs involved, then the document succeeded. The economic analyses in this "final" EIS are general sketches to accompany the concept plan. If option 2 or 3 materializes, the next supplemental EIS documents should be accompanied by clearer description of the conference center and proposed institute, including a substantial needs assessment, along with, I assume, a much deeper economic analysis. With about 310,000 square feet to redevelop, the EIS calls for about 85,000 square feet for education and interpretation. Option 2 (day only) calls for 225,000 square feet for non-for-profit institutional issues. Option 3 (day/night) discusses a 250-room and 225,000 square foot, high-end retreat and conference center, with 25,000 square feet of meeting space. The preferred day/night option is estimated to cost $169.5 million and the day option $178.1 million.

The capital cost shortfall is $159 million for option 2 and $104 million for option 3. The shortfall for the preferred option is less because the hotel/conference center contributes almost $38 million in capital to the project.

The cost estimates are in year-2000 dollars, and the estimates of profit margin and borrowing rate assume practices and rates of that period, not the current period of much more constrained private dollars. It would be difficult to imagine building these projects for those dollars in the year 2010. Furthermore, the economic growth of northern New Jersey and New York City that began during the 1990s has ended. New Jersey and New York have lost hundreds of thousands of jobs as part of the loss of 3 million jobs in the United States in the year 2008. And losses have continued. Furthermore, there is a considerable excess of hotel and conference center space in the area. Current economic realities challenge the EIS options, especially the preferred one, as there are now fewer dollars and more competition for them.

While there is little evidence of New Jersey versus New York tension in the document, it could resurface. Competition for limited resources is part of my concern. More than a dozen projects are under way to rehabilitate and reuse sites that drain into New York Harbor. I cannot precisely estimate the costs of all of these because their plans are at various stages. The Statue of Liberty avoided this frenzy of competitive redevelopment by being redeveloped and reopened in 1986 for its bicentennial. When 9/11 occurred, the Statue of Liberty was closed, and the crown was not reopened for visits until July 4, 2009.

In addition to harbor-related projects, private and government capital has been used to help build new stadiums and/or infrastructure for the Yankees and Mets (baseball), Giants and Jets (football), and Devils (hockey), and one is being considered for the Nets (basketball) – this is in addition to stadiums for minor league teams and college arenas. None of these projects may be in direct competition for capital with Ellis Island. But in the current economic climate, I suspect that some are, at least indirectly.

Governors Island might be the most obvious competition. In 2008, the NPS released a 691-page final general management plan and EIS for Governors Island (National Park Service 2008). For over two centuries, Governors Island played a vital role in the defense of New York and New Jersey. During the Revolutionary War, the Continental Army placed artillery on the island to block the British from winning the war. After independence was achieved, Fort Jay, Castle William, and South Battery were built on Governors Island. These historical fortresses are what the NPS wants to preserve and feature. After the War of 1812, Governors Island was used to hold prisoners, by the military for recruitment, and as a prison (the northeast version of Fort Leavenworth and Alcatraz). The site was enlarged to 172 acres and became the largest US Coast Guard base.

In 2001 and 2003, Presidents Clinton and Bush certified the island as a National Monument. The EIS presents Governors Island as a place to feature coastal defense and ecology of the harbor. The capital and annual operating cost of the preferred project are, respectively, $50–60 million and $11–13 million. Are there financial resources for Ellis Island and Governors Island, as well as the other projects in the harbor? Will compromises be required that satisfy no-one? Will all or some of these projects be sacrificed in the current economic climate? Will this cause the New Jersey–New York rivalry to be rekindled?

## Stakeholder reactions

The organization that will manage and sustain the Ellis Island dream will need internal cohesion. Accordingly, I read every written comment that was submitted to the NPS to attempt to assess support and concerns. The NPS created an interdisciplinary group to list and prioritize the issues, contact stakeholders, organize and conduct meetings, and obtain feedback. They relied on previous, 1995, EIS documents for the Statue of Liberty and Ellis Island. Much of the information from the older EISs is directly applicable; the September 11, 2001 terrorist attacks have required additional discussions about security and the need for a permanent bridge.

The NPS held three open meetings to scope out an EIS for the deteriorating areas. These were held in December 2000 on Ellis Island (sixteen attended), and in Trenton, New Jersey (thirteen attended) and Manhattan (fifteen attended). More than a dozen organizations expressed support for the preservation and reuse of Ellis Island during the December 2000 scoping sessions, including the New Jersey General Assembly, City of Jersey City, Liberty Science Center, New Jersey State Historic Preservation Officer, Governors Advisory Committee on the Preservation and Use of Ellis Island, Preservation New Jersey, New York Landmarks Conservancy, Liberty State Park Development Corporation, and the New Jersey Department of Health. On February 28, 2002, the NPS hosted a workshop on an Ellis Island Development Concept in New York City.

These meetings assisted in the preparation of the concept land-environmental impact statement of June 2003. Prior to the preparation of the document, the NPS followed routine EIS public notification protocols. The NPS and the EPA published a Notice of Intent in the *Federal Register* to prepare the environmental impact statement. Upon completion of the draft EIS, they published a Notice of Availability when the draft document was released for public review in May 2003. Letters were sent to the New York and New Jersey State Historic Preservation Officers, as well as the Advisory Council on Historic Preservation. Press releases or articles were published in local and regional newspapers. These publications include *The New York Times*, *The Star-Ledger* (Newark), and the *Jersey Journal* (primarily Hudson County). The executive summaries for the *Development Concept Plan/Draft and Final Environmental Impact Statements* were posted on the NPS website, and the draft and final documents were made available in their entirety on the NPS planning website.

During the 60-day review period of the draft document, the NPS held two open meetings to provide additional opportunities to comment on the document. The first of these was held on July 24, 2003 at Federal Hall in Manhattan, and was attended by fourteen people. The second meeting was held on July 28, 2003, at Jersey City University in Jersey City, New Jersey; it was attended by thirty-eight people. During the period of public comment on the draft EIS, letters, cards, and emails were submitted, in addition to written and verbal statements submitted at the public meetings. I found a list of organizations that participated during the scoping process and in response to the draft EIS. These included state government and their representatives, as would be expected. However, they also included individuals representing not-for-profits and for-profits, such as the Advisory Council on Historic Preservation, Advocates for New Jersey History, Circle Line Statue of Liberty–Ellis Island Ferry, Inc., Guides Association of New York, Jersey City Office of Economic Development Liberty Science Center, Save Ellis Island Inc., National Trust for Historic Preservation, US Army Corps of Engineers, New York District, US Coast Guard, EPA, US Department of the Interior, Fish and Wildlife Service, US Department of Transportation, and the Federal Highway Administration Historic Preservation Office. Overall, it appears to me that the NPS tried to obtain feedback from the two states directly involved, and from the surrounding local community in New Jersey. Not to have done so would probably have killed this project because of the political rivalry described above.

I briefly summarize some of comments to provide a flavor of what clearly was an extremely positive response to the EIS. The tone of the letters was positive, and most of the suggestions called for fine-tuning the ideas and coordinating with other responsible parties in the region. I saw no evidence of opposition to the two alternatives, and there was no support for the no-action alternative. Beginning at federal level, the Region II office of the EPA indicated that the EPA saw no significant adverse environmental impact. However, they would review site-specific documents for specific projects, such as the proposed permanent bridge.

The National Trust for Historic Preservation supported the opposition to a pedestrian bridge. The New York State historic preservation officer opposed public access via the proposed permanent bridge, and reiterated that New York State supported public access via ferries. New York's Landmarks Conservancy strongly supported the preferred alternative, and was very pleased to see that the idea of demolishing some of the buildings for new commercial use was not included. The Mayor of Jersey City supported the preferred alternative, and called for the NPS to reconsider the idea of a permanent bridge that would be accessible by pedestrians. Across the river in lower Manhattan, the Battery Park City Authority supported the two alternatives and called for following environmentally sustainable building practices.

The National Ethnic Coalition of Organizations Foundation (with headquarters in Manhattan) praised the document and requested a meeting to discuss how the rehabilitative facilities could be used to enhance ethnic harmony.

New Jersey's Commissioner of Environmental Protection supported the proposed alternative or other alternatives that would be economically feasible and sustainable. Representing New Jersey's Environmental Review group within the New Jersey Department of Environmental Protection, a staff member provided a lengthy set of comments that, in essence, called for greater clarity about the proposals and interactions with the Department's permitting function and with the staff of adjacent Liberty State Park. For context, Liberty State Park is approximately 800 acres and over 2 million people visit it each year (see Figure 3.1).

The Liberty Science Center is a 200,000-square-foot museum located in Liberty State park in New Jersey. It offers educational programs for children. The President of the Liberty Science Center strongly supported the EIS concepts, and emphasized the need to coordinate closely with planning at Liberty State Park and Liberty Science Center (see Figure 3.1).

A group called "Friends of Liberty State Park" was extremely concerned about the continued presence of the road to the temporary bridge, calling it a "visual and physical intrusion" (National Park Service 2005, p. 230), and regarding it as in conflict with the Park's normal activities. They recommended limiting access to the bridge and tearing it down as soon as possible. The group is not opposed to a permanent bridge, as long as it doesn't cut through the park; in fact, they made a suggestion for a less intrusive location, and they do reiterate that a permanent bridge should allow pedestrians to walk to Ellis Island.

The Jersey City Department of Housing and Economic Development praised the plan and called for public access to Ellis Island via a permanent bridge. New Jersey's historic trust group called for carefully balancing public/ private partnerships with public access and public use of the island. It commended the document for calling for discounted fares and special programs to bring low-income populations to the island.

Save Ellis Island, a not-for-profit organization, supported the concepts and emphasized the need to develop a transportation plan with Liberty State Park officials, as well as infrastructure plan for Ellis Island. Among the eighteen-member Board of Directors were five local elected officials and other prominent individuals from the surrounding areas of New Jersey. They expressed some concern about the accuracy of the cost estimates presented in the EIS. A faculty member from Jersey City State University, located a short distance from Ellis Island, strongly supported the idea of an Ellis Island Institute that would remind Americans of the importance of immigration. In his comments, Senator Robert Menendez of New Jersey supported the preferred alternative and characterized the no-action alternative as "a travesty. It would be the complete loss of thousands of stories of immigration" (National Park Service 2005, p. 254).

# Interviews

John Hnedak is the Deputy Superintendent, Business Management, Planning and Development at the Statue of Liberty National Monument and Ellis Island. John was extraordinarily helpful, providing me with documents and suggestions. On July 27, 2009, I asked him questions about the project. He indicated that the EIS process was critical, and that the project as currently moving forward would not have been possible without the EIS process. The context for that statement is that, prior to the process that culminated in the year-2005 EIS, the NPS had tried to lease property on Ellis Island to a private developer under NPS-legislated authority to do so. The developer's proposal included extensive demolition of some of the historic structures as well as significant new construction. John noted that, as a result, the historic preservation communities of both New York and New Jersey voiced strong opposition to the project. Necessary compliance with national historic preservation requirements could not be achieved, and the project did not move forward. Members of Congress expressed concern over this result, and directed the NPS to continue planning for a project that benefited the public without demolitions or extensive new construction. The EIS represented the planning and environmental management tool that allowed planning to proceed. John Hnedak: "A project of this magnitude and importance is, by its nature, controversial enough to warrant the use of an EIS process to guide it. It allows us to systematically obtain input and support from the public and numerous communities of interest."

John noted that plan is moving forward, although at a slower pace because of the recession. The Laundry Hospital Outbuilding is under rehabilitation by SEI. A great deal of progress has been made on the planning for seven of the buildings to be used primarily for historic interpretation. Most of these have been stabilized. All of the programming is designed for the public. He commended the Ellis Island Institute for their high-quality designs and educational programs. The Baggage and Dormitory Building, the largest, will be stabilized by the NPS with an $8.8 million grant from the American Recovery and Rehabilitation Act. The conference center is still part of the plan. The State of New Jersey has been very supportive with dollars and political efforts. The City and State of New York have been less active. John hopes for stronger movement on fundraising in a year or two.

Elizabeth Jeffery is Vice President, Planning and Capital Projects for the Ellis Island Institute and Save Ellis Island, Inc. Trained in design and urban development, Ms Jeffery had extensive experience in physical and business planning, and in housing and urban redevelopment, before assuming her role with the private arm of this government–private partnership to restore and adaptively reuse the historic buildings on the south side of Ellis Island. I spoke with her on September 2, 2009. Ms Jeffery described Save Ellis Island as the official partner for the rehabilitation and adaptive reuse of the twenty-nine unrestored buildings on Ellis Island. She pointed to two legal documents. On January 18, 2007, Save Ellis Island and the NPS entered into an agreement to provide for SEI's fundraising, planning, construction, and programming for the currently vacant and deteriorated buildings on Ellis Island's south side, comprised primarily of the Immigrant Hospital and the Baggage and Dormitory Building. Then, on June 19, 2009, they signed a Memorandum of Intent that described progress and decisions on the steps to fully establish the Ellis Island Institute as envisioned in the EIS described in this chapter. Ms Jeffery notes that the MOI also affirms the current federal commitments to the project.

Ms Jeffery characterized the EIS as a master and land-use plan, zoning document, and an assessment of environmental effects of implementing the plan. The document, she noted, does provide details because it was already dated when written and needed to be flexible. However, it was critical because the preferred alternative described by the EIS is the Ellis Island Institute and Conference Center. Also, the EIS decided upon a permanent operations bridge, new utilities, and no significant demolition or new construction – all of which were fundamental to the reuse plan.

She traced the founding of Save Ellis Island to the 1998 Supreme Court decision granting sovereignty to New Jersey of 22.5 acres of Ellis Island. Following the court decision, former New Jersey Governor Christine Todd Whitman established a commission to study the issue and make recommendations. The Report of the New Jersey commission was formally submitted to the NPS during the EIS public process. The Commission evolved into the 501(c)(3) nonprofit organization Save Ellis Island in 1999.

Ms Jeffery was very clear about the role of the EIS. The EIS represents a "snapshot in time," she noted, and as such should not be overly proscriptive, because a great deal of analysis is required to design the best combination of land uses and activities. The EIS finished its public process in 2002 with very few objections or comments. Nevertheless, it took 5 years for it to be approved and published in the *Federal Register*. That delay was costly. Save Ellis Island was not able to obtain federal approval for the project, so fundraising was nearly impossible. In addition, several strong markets and an overall active philanthropic climate were missed. Apprehension and overly cautious attitudes at the NPS also delayed implementation. The EIS helped bring together interested parties, increased awareness, established the basic parameters for the project, and helped establish inter-governmental and inter-agency communication at the beginning. Due to its iconic stature, having public input on the planning for this site is important.

She concluded that establishing the basic plan and approving the fundamental elements of that plan were critical. For example, a major reduction of available space or a significant new requirement could have made that complex and challenging project unfeasible. Also, an EIS proactively addressed and took conclusive actions regarding the bridge and other contentious issues, which was important because, if a project were subject to constant revision and inconsistent requirements, it would be impossible to finish.

Ms Jeffery suggested that the EIS was a legal foundation for Save Ellis Island and the NPS to develop and begin implementing their ambitious plan (Save

Ellis Island 2006, 2007, 2009; see www.saveellisisland.org for updates). The *Mission, Goals and Vision* document begins with Emma Lazarus's (1883) often-read, moving poem:

> . . . "Give me your tired, your poor,
> Your huddled masses yearning to breathe free,
> The wretched refuse of your teeming shore.
> Send these, the homeless, tempest-tost to me,
> I lift my lamp beside the golden door!"
>
> (Save Ellis Island 2006, p. 1)

The document points out that almost 10% of immigrants who came through Ellis Island arrived sick:

> When 2 million people each year visit the Ellis Island Immigration Museum in the restored Registry Building to pay tribute to this immigration story, they miss the story of the *other* Ellis Island – an evocative complex of hospital buildings where the sick were treated with compassion and America's public health was protected. This is the story of the "other" Ellis Island – a story that powerfully resonates today as once again we confront global health issues and the massive migrations of people around the world. This is the story that "Save Ellis Island" is mandated to communicate and preserve, through the rehabilitation and re-use of 30 historic buildings that sit nestled in a park-like setting on the south side of Ellis Island, with a dramatic view of the Statue of Liberty and Manhattan – so near and yet so far for the immigrants who were too sick to enter the country.
>
> (*ibid.*, p. 2)

The document describes this partnership with the NPS as "one of the largest historic preservation projects in America" (*ibid.*, p. 2), and adds that the project will "further public understanding of the global migrations of peoples and the importance of health and well-being for the human community" and "demonstrates the economic and social benefit of historic preservation for civil society, and the relevance of historic places in the twenty-first century."

In light of the criticism of EISs as not collaborative, Ms Jeffery pointed to the objectives of this partnership. Save Ellis Island expects the two partners to build relationships with government agencies such as the Centers for Disease Control and Prevention, the US Public Health Service, the US Citizenship and Immigration Services, and state and local departments of public health, to raise the profile and reputation of the Institute and enable it to reach broader audiences, as will partnerships with NGOs and policy institutes such as Doctors Without Borders, the Pan American Health Organization, the Urban Institute, and the Migration Policy Institute. Partnerships with health corporations, including those in the pharmaceutical field and other medical-related businesses, will increase the Institute's visibility in the corporate world, both as a source of support and as an audience for Institute conferences and programs. Ethnic organizations will be solid partners for the Institute, including those that represent current immigrant groups and those representing smaller and lesser-known immigrant populations.

The Institute will be an educational institution in the broadest sense, and its partnerships with formal educational institutions will be paramount. The Institute has already established a robust program for school teachers in grades K–12, and will expand this partnership to engage community colleges, universities, schools of public health, graduate programs in historic preservation, and scholarly professional organizations with compatible missions. Through these strategic partnerships, the Ellis Island Institute will construct a local, regional, national and international presence and reputation for sound, quality programs, known and respected in the fields of migration and public health.

(*ibid.*, p. 12)

Perhaps because of her background in planning and design, Elizabeth Jeffery has a clear concept of the role of the EIS in a public–private partnership. The EIS is the flexible conceptual master plan that provides legal authority for action. The EIS allows the partners to move forward with a program that requires modifications that cannot be captured in a single EIS. She views partnership with the NPS as an effective way of overcoming the tendency of government bureaucracy to be overly cautious and afraid to make mistakes, and to drag its feet. Save Ellis Island has been able to raise over $40 million for restoration work and development of educational programs. Elizabeth noted that Save Ellis Island was moving more slowly than a private company might, but the EIS and government partnership provide fundamental principles and certainty on what can be done and allow the not-for-profit partner to be proactive.

My third interview was not a single conversation, but rather the summation of a number of contacts during the past five years and media reports. Senator Robert Menendez of New Jersey has been an outspoken proponent of redeveloping Ellis Island. The fact that the EIS was strongly supported provided Senator Menendez with an opportunity to attract the attention of Ken Salazar, who had been nominated as Secretary of the Interior by President Obama. The Department of the Interior has 60,000 assets and a $9 billion backlog in deferred maintenance (Hennelly 2009). Even a good, widely supported project could get lost in this backlog. At Salazar's confirmation hearing, Senator Menendez raised the issues of the Statute of Liberty and Ellis Island, and the nominee promised that he would visit. The first formal visit of the new Secretary was to the Statue of Liberty and Ellis Island.

Senator Menendez remarked:

Secretary Salazar showed that he is a man of his word by visiting the national treasures a week after he committed to me during his confirmation hearing that he would do so. It's one thing to listen to the arguments for why we should reopen these national treasures from afar, but it's another thing to be here in person, where you can fully appreciate what an unrestricted visit to these landmarks means to all of us Americans. It really has an impact on you and your sense of country. We climbed to the top of Lady Liberty's crown, where I helped make the case that it should be reopened – it's a very moving and powerful setting in which to have that discussion. On Ellis Island it was clear that the Secretary was moved by our visit to the dilapidated buildings which remain closed and the history that they hold within their walls. The Secretary must now make sure he does a thorough examination into the reopening and restoration process, and he has some further information to gather. I am optimistic that he will look for a management solution that allows Lady Liberty's crown to reopen while ensuring the safety of its visitors. I am also optimistic that he will look for the ability within his upcoming budgets to help complete the full restoration of Ellis Island.

(Save Ellis Island 2009, p. 1)

When interviewed by a radio reporter, the Secretary remarked: "As we look at the economic recovery effort for the nation President Obama and the members of Congress are looking at, certainly places like this ought to be primary candidates to be able to fund to get those jobs going" (Hennelly 2009).

## Evaluation of the five questions

### Information

I am familiar with the waters of the New Harbor from water quality studies, and have visited and studied the land forms on and surrounding Ellis Island. There are no major holes in the environmental resources database. Some of the descriptions (such as noise impacts) are thin, but the estimates seem reasonable. In fact, these are not major issues. The most serious shortcoming I found is the limited economic analyses. As someone who has worked on and critiqued economic impact and life-cycle studies, it is not unusual for me to conclude that I want more information. This economic presentation is so limited that it lacks credibility – it truly fits the expression "back-of-the-envelope calculations."

Another gap is the non-monetary values of the historical structures and artifacts on the island. The descriptions of the structures might as well have been taken from an encyclopedia. With all due respect to the NPS and the writers of this report, the descriptions are boring and the photos too limited. I understand and teach the value of scientific objectivity. However, the emotions that Ellis Island stirs for many of us are not conveyed. I would argue that scientific objectivity would have survived several well-placed quotations or stories told by people who landed on the island. These could have been woven into the presentation as side bars.

Another shortcoming is evacuation and emergency response. The two preferred options include a permanent bridge. Part of the justification is emergency response. The data offered to justify this addition are limited. Could people not be evacuated by fast-moving ferries, police boats, or other fast-moving vehicles? Could terrorists not attack via a permanent bridge, which they could then demolish? There may be good answers to these questions. But the questions were not raised in the document, which seems a surprising omission for this location – a short distance from the former World Trade Center towers.

The tone and writing are appropriate for a general audience. The evidence presented favors the preferred option. However, none of the options is dismissed out of hand, including the no-action option.

## Comprehensiveness

The document includes cultural, environmental, economic, historical, and social considerations, emphasizing the cultural/historical legacy of the site. It certainly is comprehensive in evaluating individual categories of potential impacts. I would have liked to have seen an analysis of the cumulative impacts of the Statue of Liberty, Ellis Island, and Governors Island, and of the combination of Ellis Island, Liberty State Park, and the Liberty Science Center. In general, I believe that the cumulative social and economic effects of these sites as combined packages are greater than those of any one individual project. In particular, the combination of Ellis Island, Liberty State Park, and the Liberty Science Center would likely provide greater justification for a permanent bridge than just Ellis Island.

## Coordination

The NPS contacted other federal agencies, and state and local governments. If there is a missing ingredient, it is the NPS expressing its big picture. I do not understand why the NPS plans for all the sites in the area are not provided in sufficient detail to help the reader understand how Ellis Island fits in with Governors Island and others. If there is no overall design, then that should be stated. If there is one that is detailed elsewhere, it should be described. Notably, the Governors Island EIS is more forthcoming about all the projects in the region, but not much more. This shortcoming left me to infer that this is part of the history of lack of cooperation between New York and New Jersey.

## Accessibility to other stakeholders

The NEPA process provided multiple venues to input values and suggestions at locations in Trenton, Ellis Island, and lower Manhattan, as well as through the mail. The responses were from senior elected officials, government staff, and representatives of not-for-profit organizations. Few individual citizens submitted testimony. This result has to be a slight disappointment to the NPS. A large number of letters from residents of Jersey City, teachers, and other potential visitors would have made a stronger statement to the elected officials who control the dollars.

## Fate without an EIS

There is so much support for this project from immigrant groups, Save Ellis Island and the State of New Jersey that a program would be under way without an EIS. But what kind of a program? And what kind of final objectives? It is hard to be certain that, absent the EIS process, the final land-use decisions would have benefited the public. That is, I have too often seen historically notable land uses across the United States demolished or retrofitted for profit-making endeavors such as gambling casinos, sports arenas, parking lots, and a host of others. The EIS functioned as a planning and organizing tool that gave stakeholders access to the process, and the decision-makers the legal authority to control land uses. The EIS requirement was the security blanket for the south side of the island, protecting it from profit-making abuse. The downside of this EIS is that it has taken a long time to implement the process. Arguably, without a formal EIS process, elected officials could have been persuaded to set aside the resources for the project and to worry about the impacts later. Certainly, the timing of the project has hurt fundraising. However, I suspect that the project would have run into legal challenges without the EIS process, and, as noted above, the preferred alternative could have been set aside without the process. I conclude that the NPS successfully adapted the EIS process to the needs of this widely supported, public-oriented, historical–cultural project.

# 4 Sparrows Point, Maryland: proposed liquefied natural gas facilities

## Introduction

When natural gas is cooled to below –260°F, it liquefies into liquefied natural gas (LNG), which occupies only 1/600th the volume of natural gas. This property means that LNG can be economically transported across the globe in specially designed tankers. LNG has received a great deal of public attention because of fears of leaks, explosions, and fires. For example, LNG is shipped to the United States from Trinidad and Tobago (major supplier), as well as from Algeria, Egypt, Nigeria, and about a half dozen other sources. Japan, South Korea, Taiwan, France, Spain, and other nations are much more dependent on LNG than is the USA.

The future for LNG imports is debatable. Will the United States become more like Japan and other, more LNG-dependent nations? Some assert that LNG is essential because US natural gas supplies are decreasing and LNG would boost the US economy (Greenspan 2003). Responding to Federal Reserve Chair Alan Greenspan's positive remarks about LNG, Pat Wood (2003, p. 1), then chair of the Federal Energy Regulatory Commission (FERC), welcomed them as "underscoring the economic importance of abundant, reasonably-priced natural gas."

In April 2009, The US Energy Information Administration (EIA) reported that it expected a sharp increase in LNG imports, from 352 billon cubic feet in 2008 to over 500 in 2009 (LNGpedia 2009). Long term, the EIA (2009) estimated that LNG could increase to 2.8 trillion cubic feet in 2030, and could account for over 20% of natural gas imports into the United States. There are ten operating facilities in the United States (a substantial increase from only five a few years ago), two in Mexico, and one in Canada. Over thirty sites were being considered at the time of the Sparrows Point LNG application.

While some favor more LNG imports, others argue that LNG–natural gas is another fossil fuel that we should use less of, not more. Instead, they assert we should be investing in conservation and alternatives such as solar and wind.

Given the relatively limited role of LNG in the United States at this time, why would I pick an LNG facility? I had three reasons. There has been criticism that federal agencies do not cooperate with each other and their state counterparts on EISs. The Sparrows Point EIS was written by the FERC, but had input and review from multiple others, and the proposed project, if built, will be monitored by multiple agencies. So one reason for choosing an LNG proposal is that presumably it demonstrates genuine cooperation among federal agencies. Second, I wanted a case study that had both water and land impacts. The proposed project consists of a major fixed facility project and two transportation projects. Organizationally and geographically, this LNG facility resembles a spider's web, reaching long distances across oceans and short distances on land. Third, I wanted a case study that was the product of a regulatory agency. I have had experiences with the Atomic Energy Commission (no longer in existence) and with the Nuclear Regulatory Commission (NRC) regarding siting of nuclear power plants. I did not want to work on an EIS concerning nuclear power plant siting because of this personal history, which included developing some recommendations that were incorporated into NRC siting guidelines. An LNG siting case, I expected, would be similar in some respects to the NRC process, and would allow me to illustrate the special format and tone of regulatory-driven EISs.

## The organization web: regulating and managing a hazardous substance

The EISs in Chapters 2 and 3 had unambiguous messages. They were not advocacy documents *per se*, but not far from advocacy. The Driscoll Expressway EIS promoted the road; the Hudson–Bergen line EIS favors the line, stopping to evaluate the options and listening to stakeholders, which the Driscoll EIS did not. The Ellis Island EIS unabashedly asserts the importance of the island, and focuses on its unique role in American history. I had the feeling that I would be unpatriotic to oppose the idea of rehabilitating the south side of Ellis Island.

In strong contrast, this chapter illustrates how regulatory agencies are required to balance on a narrow rope when they write an EIS. They assemble, evaluate, and present masses of technical detail. Typically, so much information is presented that critics argue that regulatory agencies obscure key information in a sea of details. A regulatory EIS tone is flat, in essence emotionless. The typical regulatory-based EIS message is: we may permit you to build and operate this facility, but you must respond to our recommendations and conform to all our regulations during planning, construction, tests, operation, and closure. Applicants are not discouraged, and perhaps they are encouraged, by the expectation of ultimate approval if they can work through the regulatory maze. Approval of an applicant's proposal is a major victory, but is only temporary as the requirements include ongoing monitoring, surveillance, and reporting. If there is a tone in a regulatory EIS, it is buried in recommendations offered by the regulatory agency. I have deliberately highlighted and directly quoted some recommendations in the EIS's language rather than succumb to the temptation to summarize them.

The staff of the FERC, an independent body within the US Department of Energy, wrote the draft EIS for the proposed Sparrows Point LNG Terminal and the Mid-Atlantic Express Pipeline Project. The US Army Corps of Engineers, US Coast Guard, US EPA, and Pennsylvania Department of Conservation and Natural Resources cooperated.

FERC has authority for siting of onshore LNG facilities under two sections of the Natural Gas Act of 1938. The Energy Policy Act of 2005 (P.L. 109-58) clarified any ambiguity in that authority by granting FERC "exclusive" authority (see Parfomak and Vann 2008). FERC is to write regulations for pre-filing of LNG terminal applications and to consult with appropriate state agencies regarding safety issues. FERC is the designated "lead agency" for co-ordinating among all federal agencies and for all environmental compliance, including preparation of an EIS. Applicants are to provide FERC with reports that analyze a full spectrum of environmental, public health, and socio-economic considerations (see Table 4.1 on p. 92). FERC has ongoing responsibilities, and has a branch to monitor existing LNG facilities.

The US Coast Guard is responsible for determining the suitability of the waterway for LNG marine traffic. The Coast Guard must issue a Letter of Recommendation regarding suitability. As part of the US Department of Homeland Security, the Coast Guard also is central to assessing and providing security at LNG sites (GAO 2007; Hurst 2008).

The National Fire Protection Association (NFPA) is a not-for-profit organization that wrote standards for LNG terminals in 1967, and these have been revised ten times. The NFPA standards are incorporated into LNG terminal safety regulations.

Under the Pipeline Safety Act of 1979 (P.L. 96-129, amended three times), the US Department of Transportation (DOT) sets safety standards for onshore LNG facilities. This mandate includes minimum standards for siting, design, construction, and operation of LNG facilities. In 1980, the DOT established a set of safety standards, *Liquefied Natural Gas Facilities: Federal Safety Standards*, under Title 49, CFR, Part 193. These standards apply to the construction, operation, and maintenance of onshore LNG facilities. But FERC, not DOT, approves specific LNG siting proposals. FERC and DOT have an inter-agency agreement to determine their complementary responsibilities.

States can prepare their own policies; however, these must be consistent with federal regulations. The key federal–state conflict is about siting (Dweck *et al.*

2006). The Sparrows Point proposal is a clear case of FERC's authority being challenged by state and local government, so far unsuccessfully. At the time when this chapter was written, there were seven Congressional bills to reduce FERC's exclusive decision-making powers regarding LNG siting. The US Court of Appeals has responsibility for reviewing FERC's decisions (see below for the Sparrows Point case). The President and Congress do not review FERC's siting decisions.

The requirement for a complicated, multi-agency agreement is the product of LNG's history. LNG's temperature is –260°F, and it can disperse as a vapor. As a liquid, it can cause freeze burns and, depending on the length of exposure, even more serious injury. If LNG comes into contact with materials while in its cryogenic (very cold) state, exposed material will be subjected to extreme thermal stress. Materials not designed for contact with extreme cold likely will become brittle, fracture, or lose strength. This reaction is not unique to LNG. Liquid oxygen, nitrogen, helium, and other cryogenic gases share similar properties.

LNG vaporizes very quickly when exposed to the ambient environment. Each cubic foot of liquid becomes 620–630 cubic feet of natural gas. If a large quantity of LNG is spilled in the presence of an ignition source, a so-called "pool fire" will result, with high levels of heat in the area surrounding the LNG pool. A large quantity of LNG spilled without a nearby ignition source will vaporize into a cloud and follow the prevailing wind until it either disperses below the flammable limits (hopefully) or encounters an ignition source, creating a serious thermal hazard. If LNG spills into water, a rapid phase transition (RPT) occurs as the water immediately changes from liquid to gas. An RPT can lead to large jumps in pressure and to damage of materials, including vessels.

Taking a step back from LNG, methane is an asphyxiate and is flammable. However, methane is a familiar hazard, and people have become acclimated to being around it and using it. LNG is a relatively unknown hazard, especially in the United States.

LNG-related risks are more than theoretical. In 1944, there was fire at an LNG facility in Cleveland, Ohio, which has been blamed on the use of materials not appropriate for cryogenic temperatures and the lack of spill impoundments. A resulting explosion and fire killed 128 people. In 1979, an accident occurred at the Cove Point LNG facility in Lusby, Maryland (one of the alternative sites considered in the Draft Environmental Impact Statement, DEIS). A pump seal failed, resulting in gas vapors entering an electrical conduit. A worker turned off a circuit breaker, igniting the gas, killing a worker, and damaging the facility. Reports have examined more than a dozen other events and used simulation models to understand LNG risks (FERC 2008a; Parfomak and Vann 2008; Rabaska 2008; California Energy Commission undated).

The risks have led to changes in the US national fire codes for LNG facilities. On January 19, 2004, an explosion occurred at Sonatrach Skikda LNG liquefaction facility in Algeria, which killed twenty-seven workers and injured fifty-six. This was caused by a leak in equipment that is not used in US systems. The fact that the event cascaded through different components of the system is troubling (FERC 2008a).

As a result of this history, and the land–sea interface of LNG, three federal agencies share regulatory authority over the siting, design, construction, and operation of LNG-import terminals. The FERC is the lead federal agency responsible for the preparation of analyses required under the NEPA for impacts associated with terminal construction and operation.

In February 2004, FERC, the Coast Guard, and DOT signed an Inter-agency Agreement to make sure that they work in a coordinated manner to manage safety and security issues at LNG-import terminals (see below for more detail). The FERC coordinates its pre-authorization review of the proposal with the Coast Guard and the DOT to ensure appropriate safety and security reviews.

## Proposed preferred project: an engineered web

A detailed summary of the proposed project is possible because the technologies are not that complicated, certainly less complicated than those presented in Chapters 5 and 6. Yet presenting even a minimally comprehensive description would require hundreds of pages of text and diagrams. Embedded in this EIS are the equivalent of three separate projects that arguably merit an EIS:

- LNG tanker trip to land berths
- facilities to unload the LNG and convert it back into gas
- transmission through a new pipeline into the gas network.

The DEIS (FERC 2008a, p. 4–172) estimated the cost of the project at $815 million.

To avoid overloading this text with details, I have been selective, even more so about the alternatives than the preferred action. I chose to focus on those elements that make this project challenging for the proposed location – the LNG tankers and dockside facilities, rather than transport of the LNG via tanker and natural gas through 88 miles of pipelines.

A caveat about this choice is that digging up open spaces and people's yards and crossing urban areas, rivers, and wetlands with a 30-inch pipeline may be more distressing to people than the LNG tanker trip and terminal facilities. But pipeline construction projects are common, so I have not described impact elements such as demolition and construction. Nor have I described methods for revegetating and restoring areas disturbed by laying new pipe and for storage of construction materials. I have not focused on loss of land for the terminal facilities or pipelines. I have not discussed using ballast on unloaded LNG ships. Nor have I emphasized operation and maintenance requirements for the system that are required by federal and state regulations and the American Society of Mechanical Engineers Standards, nor storage tank construction requirements. Finally, despite some recent publicity about gas line explosions, I have not discussed the safety issues associated with building and operating a new gas pipeline. None of these is a trivial issue, and each should be discussed as part of the EIS process for this or any similar facility.

With these caveats noted, AES and Mid-Atlantic Express proposed to build and operate the Sparrows Point LNG Terminal in Baltimore County, Maryland to import LNG, store, vaporize and transport about 1.5 billion cubic feet of natural gas per day (Bcf/d).

## Getting to the unloading sites

LNG tankers would come to Sparrows Point from sources such as Algeria, Australia, Brunei, Indonesia, Malaysia, Nigeria, Oman, Qatar, Trinidad and Tobago, and United Arab Emirates. If you ever see an LNG tanker, you will not forget it (Figure 4.1). Newer LNG tankers are approximately 900 feet long and about 140 feet wide, and have characteristic ball-shaped tanks on top. An LNG tanker is about the size of a modern aircraft carrier – they are enormous. By comparison, the famous Liberty Ships of the Second World War were about half as long and half as wide. Construction of LNG tankers is regulated so that tankers can carry LNG on long voyages. LNG tankers, unlike many oil and chemical product tankers, have two hulls, designed to hold their liquid cargo. They contain engineered systems to safely contain liquids stored at temperatures of –260°F.

LNG tankers headed for Sparrows Point would enter the Chesapeake Bay and travel to the unloading dock about 164 nautical miles away (Figure 4.2). The channel's depth varies: some of it is dredged for large vessels and other parts have naturally deep channels. Tugboats would meet the LNG vessel and escort it along the Brewerton Channel and into the Sparrows Point Shipyard

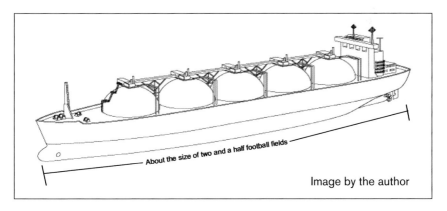

About the size of two and a half football fields

Image by the author

**Figure 4.1**

LNG tanker

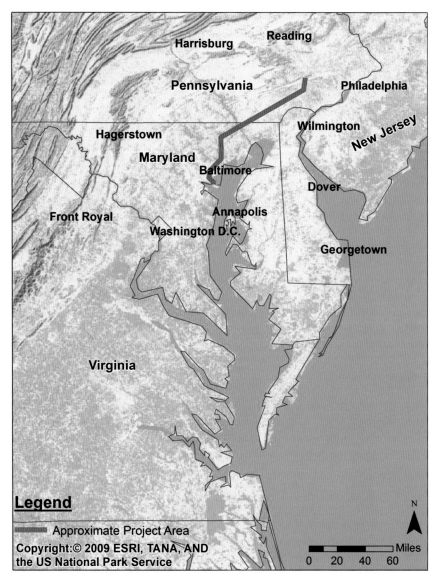

Legend

▬▬▬ Approximate Project Area
Copyright:© 2009 ESRI, TANA, AND
the US National Park Service

N

Miles
0    20    40    60

**Figure 4.2**

Sparrows Point LNG project, general project location

Channel (Figures 4.3 and 4.4). The Sparrows Point Shipyard Channel would need to be widened and deepened to accommodate the LNG vessels.

The US Coast Guard is required to assess the suitability of the Chesapeake Bay and associated water bodies for the LNG ships, and it must issue a formal letter of recommendation for the facility. LNG tankers that arrive at Sparrows

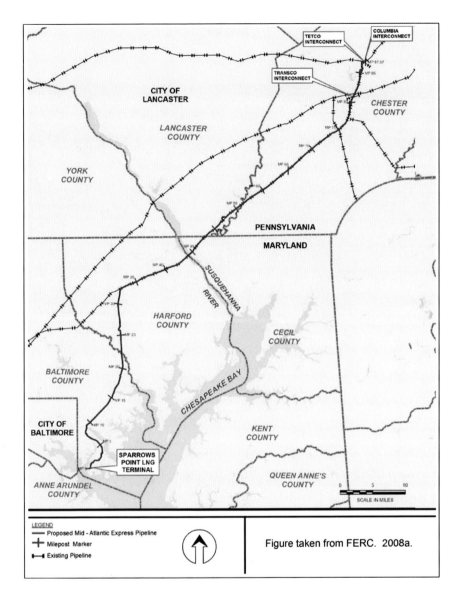

**Figure 4.3**

Mid-Atlantic Express Pipeline project

Point would be operated so that their maximum arrival draft (loaded) would not exceed 40.5 feet. The DEIS notes that the approach to the berths, the berths, and turning basin would be dredged so that the water depth is at least 45 feet at mean lower low water (MLLW) for keel clearance at all tidal stages.

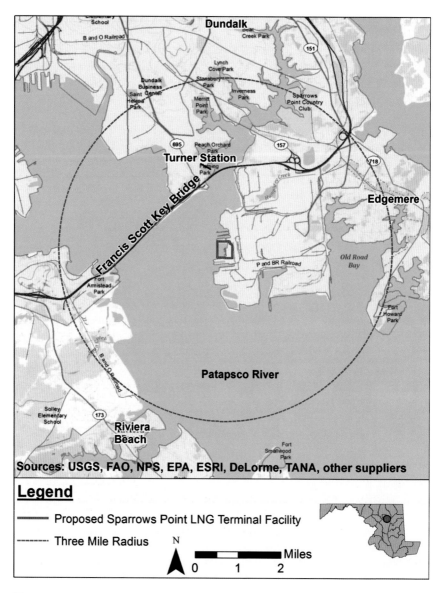

Sources: USGS, FAO, NPS, EPA, ESRI, DeLorme, TANA, other suppliers

## Legend

━━━━━ Proposed Sparrows Point LNG Terminal Facility

- - - - - Three Mile Radius

N

Miles
0    1    2

**Figure 4.4**

Proposed Sparrows Point LNG terminal and surrounding communities

## Vessel operation

AES, one of the applicants, indicated that the LNG vessels would be traveling at 15–18 knots ("sea speed") in the North Atlantic and in the Chesapeake Bay entrance and lower Chesapeake Bay. Once they move north up the Bay, the LNG tankers, tugs, and any other vessels would run far from shore for the majority of the route at speeds of 12–18 knots. The vessels would reduce speed to 12–15 knots past the Chesapeake Bay Bridge–Tunnel, and would run at 12–14 knots passing under the Bay Bridge. Within the Craighill Channel, the vessels would travel at about 10 knots, and along the Brewerton Channel vessels would slow to about 5 knots. The vessels would have a Maryland Pilot in the Precautionary Zone of Chesapeake Bay, and a docking pilot at the Cut-Off Channel before entering Brewerton Channel.

For context, the document reports that traffic into Baltimore Harbor is about 30% less than historical volumes. The DEIS indicates a decline in the number of ships arriving into the Port of Baltimore from 4033 in 1975 to 2119 in 2005. However, it adds that the size of the ships and the associated increase in value of the cargo delivered to the Port of Baltimore has increased. Furthermore, a passenger cruise terminal was established in the harbor in the Spring of 2006 to serve approximately twenty-eight cruise-ship visits to Baltimore between May and November. A 500-yard security zone has been established for cruise ships, as well as LNG ships. Cruise ships typically come in early in the morning and leave early in the evening on Fridays and Sundays. Overall, Chesapeake Bay ship traffic attributable to the proposed Sparrows Point LNG vessels would be about 7% (150 of 2100) of the Port of Baltimore traffic.

When a ship is loaded with flammable material, the focus is drawn to that hazard. Yet it would be an error not to also focus on dredging and the location of dredged materials. I have repeatedly seen dredging become a contentious issue in the United States. Construction of the LNG terminal requires widening and deepening the channel up to the existing Brewerton Channel and the turning basin offshore of the terminal site to accommodate the tankers. The report estimates that about 3.7 million cubic yards of dredged material from an approximately 118-acre area in the Patapsco River would be generated in order to meet the channel and turning basin design depth of 45 feet below MLLW (Figure 4.4). Dredging stirs up materials at the bottom of the channel, including toxins, causing problems for fish and other species. FERC requested comments from agencies, the applicant, and individuals on which dredging method is most appropriate.

AES (FERC 2008a) proposed to recycle the dredged material for the following purposes: abandoned mine land and quarry reclamation; brownfield redevelopment; landfill capping and closure; alternate grading materials; low-permeability cap layer in lieu of geo-membrane systems; manufactured topsoil; general structural and non-structural fill for commercial/industrial development; and bulk construction fill, including site grading material and highway embankments.

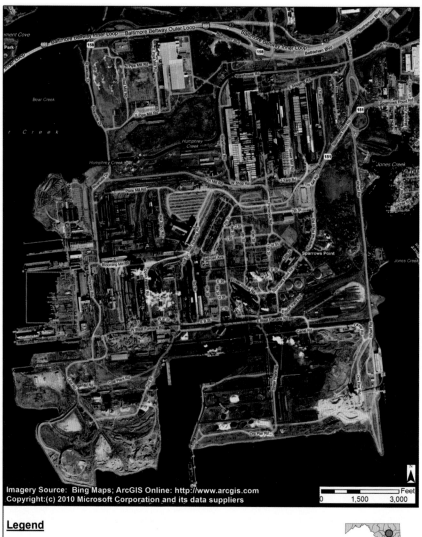

Imagery Source: Bing Maps; ArcGIS Online: http://www.arcgis.com
Copyright:(c) 2010 Microsoft Corporation and its data suppliers

Feet
0   1,500   3,000

**Legend**

——— Proposed Sparrows Point LNG Terminal Facility

**Figure 4.5**

Aerial photo of proposed Sparrows Point LNG terminal location and surrounding communities

The author is familiar with locations where dredged materials have been successfully applied; in several of these locations, dewatered, dredged materials were good applications, and indeed environmental groups supported several applications. Assuming dredging lasting two years, AES anticipates using ten to fourteen 1500–3500 cubic yard work scows to transport the initial dredged material to the processing facility. These vessels would be sealed and watertight in order to avoid any release of dredged material back into the water. The dredged sediments would be dewatered immediately or would be placed in tanks to allow settlement.

This is a large amount of dredged material, and transporting it could draw the ire of neighbors who live near sites where rail cars or trucks would pass nearby. For example, AES estimates approximately 220 truck trips a day hauling the material. Public perception, I feel certain, would be a major issue, and there could be an occasional spill that would exacerbate local reaction. In the event these preferred options for dredged materials are not viable, AES cited areas managed by Waste Management and Allied Waste Services in Virginia as alternative locations for dredged materials, potentially leading to public outrage in those areas, I suspect.

## Unloading, converting and sending out the gas

The Sparrows Point DEIS proposal describes a facility with two berths for unloading LNG ships with capacities of 125,000 to 217,000 m$^3$. LNG tankers would arrive about 120–150 days a year (two or three times a week). The tankers would unload at two separate LNG berths. One, south of the unloading dock, would be the primary unloading berth, and the northern site would be the auxiliary facility. Both of these berths would use a platform to be built on top of an existing concrete, pile-supported pier. Each berth would have three 16-inch liquid unloading arms. LNG would be unloaded from a tanker at a rate of 12,500 m$^3$ per hour into LNG storage tanks with a 32-inch-diameter unloading pipeline. This unloading pipeline would be maintained at cryogenic conditions. These facilities would be required to be designed to meet US DOT standards and requirements of the NFPA for LNG.

The unloaded LNG would go into three full-containment LNG storage tanks, each with a nominal working volume of about 160,000 m$^3$ (a total of 480,000 m$^3$) and with the capacity for full containment. Full containment means that there is a primary inner containment and a secondary outer containment. The tanks are designed and constructed so that each containment is able independently to contain the full volume of LNG. To provide a sense of magnitude of the tanks, the outside diameter of the outer containment would be approximately 270 feet at the base of each tank, and the height of each tank would be 170 feet. Each insulated tank would be designed to store a net volume of 160,000 m$^3$ of LNG at cryogenic temperatures and a maximum internal pressure of 4.3 pounds per square inch gauge.

Once unloaded, the proposal calls for two LNG vaporization systems at the terminal. One is a "sendout" system under high pressure. The other is an intermediate-pressure fuel gas system. The high-pressure system would generate the natural gas leaving the Sparrows Point LNG facility. The intermediate-pressure system generates the intermediate-pressure fuel gas for supply either to the low-pressure fuel gas system or to a possible power plant combustion turbine at the site (a power plant was an option in the proposal). The LNG vaporizers, in essence, would be vertical shell-and-tube heat exchangers, with LNG flowing on the tube side and heat-transfer fluid flowing on the shell side. The heat-transfer fluid heating system provides heat to vaporize the baseload natural gas sent from the facility.

In addition to these elements, the report notes the following as additional requirements or possible add-ons. AES is considering building a combined cycle co-generation power plant at the site. It would use natural gas to produce about 300 MW of electric power. In this scenario, the excess heat of the power plant would be used to vaporize the LNG at the project terminal. The commercial viability of this option was not determined when this DEIS was prepared. Lastly, on-site processing will require support facilities, such as a main control room, and security and administrative offices.

## *Moving the natural gas to key intersecting pipelines*

To move the natural gas, Mid-Atlantic Express proposed to construct and operate a 30-inch-diameter natural gas pipeline to connect to three interstate natural gas pipelines near Eagle, Pennsylvania (Columbia Gas Transmission Corporation, Transcontinental Gas Pipe Line Corporation, and Texas Eastern Transmission Corporation). The pipeline would run about 48 miles in Baltimore, Harford, and Cecil Counties in Maryland, and 40 miles in Lancaster and Chester Counties in Pennsylvania, near the suburbs of Philadelphia. Large pipelines criss-cross the United States and, like many potential hazards, are ignored until an event occurs – then all become suspect.

A final set of requirements is for so-called "pig launching and receiving facilities" at the beginning and ending of the proposed pipeline. These in essence monitor and clean the lines. The report also mentions the need for nine mainline valves, and three metering and regulation stations at the intersection locations at the end of the pipeline.

# Alternatives

The EIS reviewed multiple alternatives to the proposed action, including the no-action and postponed-action alternatives, Coast Guard route alternatives, LNG technology modifications, LNG terminal site options, and pipeline system and route alternatives. The DEIS evaluated four major route alternatives and thirteen route variations.

The document asserts that the no-action alternative would deprive the market area of the fuel, and this loss would not be made up by solar, wind, or other renewable energy sources. The evidence and arguments on economic impacts are the weakest I found in the DEIS. They are troublesome because, in the author's experience, citizens and local groups are usually aware of the economic issues involving options and expect a thorough exploration (see below for a more detailed discussion). With regard to alternative locations, the DEIS reviewed one site in the lower Delaware Basin (Logan Township, New Jersey); another proposed by the Philadelphia Gas Works; and a third in the Chesapeake at Cove Point (an expansion at an existing approved LNG site). Also, within the Chesapeake, the report examined six other suggested locations.

The DEIS describes the Chesapeake as the only water body (along with the Atlantic Ocean) that has sufficient depth for these tankers and at the same time is well located for reaching the proposed natural gas market. The Sparrows Point site itself is characterized as an industrial site far from residential areas. It certainly is an industrial site (former steel mill), but some who live nearby to the north would disagree with the characterization of the site as remote (see Figure 4.4).

Multiple options were reviewed for dredging and disposing of the dredged materials. Reusing the dredged materials was the preferred option, which makes sense, assuming the quality of the dredged material is suitable and there are enough locations for it.

The DEIS reviewed four terminal design options both on- and offshore, and options for converting the fuel from LNG back to natural gas. More than two dozen major and minor route options were reviewed for getting the natural gas to the existing major gas lines. Ultimately, the preferred option is supported with conditions.

## Environmental Impact Assessment: some illustrations

Table 4.1 lists the multiple impacts that the DEIS assessed. It is not feasible to review every category, so I have focused on several, and almost exclusively on the recommendations in order to illustrate the tone of regulatory EISs. As noted earlier, I directly quote sections from the DEIS specifically to illustrate the language. The first impact issue is reliability and safety, which commands the largest section of the report and is of greatest concern to the surrounding communities. What is presented below is not comprehensive, but rather offers snapshots that illustrate how regulatory EISs are presented.

### Reliability and safety

This DEIS, like most written about projects that include major hazards, relies on federal, state, and local rules and regulations to manage each part of the project. For example, the LNG terminal facilities would be sited, designed, constructed, operated, and maintained in compliance with the federal siting

**Table 4.1** Impacts considered by Sparrows Point liquefied natural gas EIS

| Topic | Impacts |
|---|---|
| Geological resources | Physiologic and geologic setting<br>Other natural hazards<br>Paleontological natural hazards |
| Soils | LNG terminal site |
| Water resources | Groundwater<br>Surface water |
| Wetlands | Regulatory permits<br>Wetland types impacted by the proposed project<br>Potential impacts of the mid-Atlantic express pipeline to wetlands<br>Wetlands construction and maintenance procedures and the aquatic resources mitigation plan |
| Vegetation | Vegetation resources<br>Vegetation management plan<br>Noxious weeds<br>Vegetation conclusions |
| Terrestrial and aquatic species | Terrestrial species<br>Aquatic species<br>Essential fish habitat |
| Threatened, endangered, and other special status species | Federally listed threatened and endangered species<br>Federally listed species on the marine transit route<br>State-listed threatened and endangered species and other species of concern |
| Land use, recreation, and visual resources | Land use<br>Existing and planned residences and developments<br>Coastal zone management<br>Hazardous waste sites<br>Recreation and public interest areas<br>Visual resources |
| Socioeconomics | Population, economy, and employment<br>Housing<br>Public services<br>Transportation and vehicle traffic<br>Property values<br>Tax revenues<br>Environmental justice |
| Cultural resources | Results of cultural resource surveys<br>Native Americans<br>Unanticipated discoveries<br>Compliance with the NHPA (Section 106) |

**Table 4.1** Continued

| Topic | Impacts |
| --- | --- |
| Air quality and noise | Air quality |
| | Noise |
| Reliability and safety | LNG hazard |
| | Front-end engineering design and review |
| | Storage and retention systems |
| | Siting requirements |
| | LNG vessel safety |
| | Emergency response and evaluation planning |
| | Conclusions on LNG vessel safety |
| | Terrorism and security issues |
| | Pipeline safety standards |
| | Pipeline accident data |
| | Impact on public safety |
| Cumulative impacts | (repeat of all twelve topics above) |

and design requirements for LNG facilities. The authors note that they cannot assess the cumulative risk for an intentional attack on the LNG facility. They add that the likelihood of ship accidents is likely to rise as international vessel traffic increases. These two points are important, and illustrate the difficulty of preparing this kind of EIS until the federal government policy-makers issue regulations and/or guidelines, in other words, choose to make a policy decision that applies to all sites, not just to one. The author has seen similar issues arise in the siting of nuclear power-related facilities.

The report emphasizes that, in order to reduce the likelihood of an accident with an LNG tanker, each LNG carrier would be under pilot control from near the mouth of the Chesapeake Bay, and have a tug escort for the last six miles. In addition, the Coast Guard would also constrain moving vessels around LNG ships. They also would have in place an emergency response plan to minimize risks. This means a plan that has the capacity to coordinate among local emergency planning groups, fire departments, state and local law enforcement, the Coast Guard, and other appropriate federal agencies, to be used in the event of an incident. At the time of this report, no such organization existed to develop this plan, although the report notes several times that the Coast Guard will require such a plan and organizational capacity.

I focus here on FERC's recommendations regarding safety. Here are some recommendations that illustrate the tone of these regulatory EISs:

Complete plan drawings and a list of the hazard detection equipment should be filed prior to initial site preparation. The list should include the instrument tag number, type and location, alarm locations, and shutdown functions of the proposed hazard detection equipment. Plan drawings should clearly show the location of all detection equipment.

AES should provide a technical review of its proposed facility that:

a. identifies all combustion/ventilation air intake equipment and the distances to any possible hydrocarbon release (LNG, flammable refrigerants, flammable liquids and flammable gases); and
b. demonstrates that these areas are adequately covered by hazard detection devices and indicates how these devices would isolate or shutdown any combustion equipment whose continued operation could add to or sustain an emergency. AES should file this review prior to initial site preparation.

(FERC 2008a, pp. 4-229 and 4-230)

With regard to operations, the DEIS states the following:

The facility should be subject to regular FERC staff technical reviews and site inspections on at least an annual basis or more frequently as circumstances indicate. Prior to each FERC staff technical review and site inspection, AES should respond to a specific data request including information relating to possible design and operating conditions that may have been imposed by other agencies or organizations. Up-to-date detailed piping and instrumentation diagrams reflecting facility modifications and provision of other pertinent information not included in the semi-annual reports described below, including facility events that have taken place since the previously submitted semi-annual report, should be submitted.

Significant non-scheduled events, including safety-related incidents (i.e., LNG or natural gas releases, fires, explosions, mechanical failures, unusual over pressurization, and major injuries) and security related incidents (i.e., attempts to enter site, suspicious activities) should be reported to FERC staff. In the event an abnormality is of significant magnitude to threaten public or employee safety, cause significant property damage, or interrupt service, notification should be made immediately, without unduly interfering with any necessary or appropriate emergency repair, alarm, or other emergency procedure. In all instances, notification should be made to Commission staff within 24 hours. This notification practice should be incorporated into the LNG facility's emergency plan. Examples of reportable LNG-related incidents include:

a. fire;
b. explosion;
c. estimated property damage of $50,000 or more;
d. death or personal injury necessitating in-patient hospitalization;

e.  free flow of LNG that results in pooling;
f.  unintended movement or abnormal loading by environmental causes, such as an earthquake, landslide, or flood that impairs the serviceability, structural integrity, or reliability of an LNG facility that contains, controls, or processes gas or LNG;
g.  any crack or other material defect that impairs the structural integrity or reliability of an LNG facility that contains, controls, or processes gas or LNG;
h.  any malfunction or operating error that causes the pressure of a pipeline or LNG facility that contains or processes gas or LNG to rise above its maximum allowable operating pressure (or working pressure for LNG facilities) plus the build-up allowed for operation of pressure limiting or control devices;
i.  a leak in an LNG facility that contains or processes gas or LNG that constitutes an emergency;
j.  inner tank leakage, ineffective insulation, or frost heave that impairs the structural integrity of an LNG storage tank;
k.  any condition that could lead to a hazard and cause a 20% reduction in operating pressure or shutdown of operation of a pipeline or an LNG facility;
l.  safety-related incidents to LNG marine traffic at or en route to and from the LNG facility; or
m.  an event that is significant in the judgment of the operator and/or management even though it did not meet the above criteria or the guidelines set forth in an LNG facility's incident management plan.

In the event of an incident, the Director of OEP has delegated authority to take whatever steps are necessary to ensure operational reliability and to protect human life, health, property or the environment, including authority to direct the LNG facility to cease operations. Following the initial company notification, Commission staff would determine the need for an on-site inspection by Commission staff, and the timing of an initial incident report (normally within 10 days) and follow-up reports.

(ibid., pp. 4-232 and 4-233)

The following sections present concerns and then requests regarding the presence of two berths for tankers; the potential for a release from a tanker; requirements for an emergency response plan that is paid for by the companies, not the local governments; the threat of terrorism; air quality and noise; and many more. Reading this document, and others like it, recalls the expression "death by a thousand cuts." Except in this case it is more likely to be life by responding to all the requests and following the rules. I cannot possibly do justice to the remarkable amount of detail provided in this illustrative EIS. As noted above, I have picked several more examples that illustrate a variety of impacts.

## Impact on aquatic systems

Resuspension of toxins and sediments, and deposition of dredged materials, are major concerns wherever there is dredging. Reading between the lines is sometimes helpful in interpreting the message of a regulatory body. In their review of dredging for the proposed LNG facility, the position of the authors is that similar dredging projects have been permitted elsewhere. The message is that dredging is a temporary annoyance, not a significant impact. The DEIS (*ibid.*, p. 4-55) acknowledges that dredging was raised by numerous commentators and agencies, and recommends that:

> Prior to the end of the DEIS comment period, AES should file with the Secretary a comprehensive Dredged Material Placement Plan. This plan should address:
>
> a. where the PDM is going;
> b. the capacity of the temporary placement areas onsite;
> c. the daily takeaway capacity for the PDM;
> d. how many daily truck trips would be necessary to haul the PDM, the impacts of those trucks on the traffic in the area, and the probable routes the trucks would take; and
> e. a contingency plan for the PDM after it is processed should there be no buyers.
>
> (*ibid.*, p. 4-55)

Furthermore, it is always possible that the dredged material may not be suitable for fill. Hence, as expected, the report calls for additional information and review.

Thousands of pipelines have crossed rivers and sometimes seriously disrupted ecosystems. The impacts can affect fish because of disturbance to the environment, sedimentation and turbidity, destruction of stream bank cover, and introduction of water pollutants. Fish can be trapped by construction activities. Yet the DEIS notes that the impact of construction on fish and other aquatic systems typically is short-term and geographically localized. A major concern is spawning fish. Therefore, the DEIS recommends consultation with the appropriate federal and state agencies to protect spawning fishes in sensitive water bodies. The applicant is required to file a report summarizing those consultations.

Despite setting forth concerns and recommendations, the DEIS concludes that adverse impacts would be "minimal." This is because the species of concern are considered by the analysts to be highly mobile and would not be impacted by ship movements along potential transit routes.

With regard to dredging, the authors conclude that the key species would be able to feed and not be seriously affected by the construction projects. The site plan, the authors assert, would minimize impacts from sound and pressure waves resulting from pile-driving. Mid-Atlantic Express would use best management practices to limit aquatic impacts from construction of the proposed pipeline. Hence, the overall conclusion is that the project would not cause notable adverse effects to species of concern or to spawning of important species of federally managed finfish.

## Visual impact

The terminal and berths would be visible from the nearest community, about 1.2 miles away (see "Environmental justice" below). Yet the terminal and Mid-Atlantic Express Pipeline projects would not cross any national parks or forests, wilderness areas, wildlife refuges, waterfowl production areas; recreational, scenic or historic trails designated through the National Trails System Act of 1968; federally designated natural, recreational or scenic areas; registered natural landmarks; or national wild and scenic rivers. Several schools, churches, local parks, camp grounds, recreation areas, and a golf course would be crossed by the proposed pipeline route. But the DEIS considers these crossing-related impacts to be temporary impacts and not significant.

Removal of vegetation and failure to restore landscape annoys the public (see light rail discussion in Chapter 2). When a long pipeline is proposed, a public response is a certainty. Hence the FERC offered a series of recommendations, including that:

> Prior to the end of the DEIS comment period, for all residences located within 50 feet of the construction work area, Mid-Atlantic Express should commit to:
>
> a. not remove mature trees and landscaping within the edge of the construction work area, unless necessary for safe operation of construction equipment;
> b. immediately after backfilling the trench, restore all lawn areas and landscaping within the construction work area consistent with the requirements of the Plan;
> c. fence the edge of the construction work area adjacent to the residence for a distance of 100 feet on either side of the residence to ensure that construction equipment and materials, including the spoil pile, remain within the construction work area;
> d. try to maintain a minimum distance of 25 feet between the residence and the edge of the construction work area; and furthermore,
> e. for any residence closer than 25 feet to the construction work area file a site-specific plan with the Secretary prior to the end of the DEIS comment period.
>
> (*ibid.*, p. 4–142)

## Socioeconomic impacts

Baltimore County, Maryland is the most densely populated county in the Baltimore Metropolitan region (approximately 1260 people per square mile). The other counties that would host the proposed pipeline have population densities ranging from 247 (Cecil County) to 573 (Chester County) people per square mile. The gross density is deceiving because most of the land that would be crossed by the pipeline is agricultural, which means that Mid-Atlantic Express would request the landowner to remove the crop, paying them fair market value for crop loss.

According to the document, population impacts resulting from the Sparrows Point project would be minimal, estimated at 420 workers recruited to build the facilities (the others would be recruited locally). Permanent employees are estimated at 75 people, which is minimal for large metropolitan region, but enough to draw some support from some labor groups for the project.

There was an environmental justice concern expressed in the area immediately north of the proposed site. Public concerns were reported in the DEIS (*ibid.*, p. 4-184) as follows:

> the Project discriminates against the African American community at Turner Station; areas affected would be burdened by the pipeline and would not receive benefits; handling of potentially contaminated sediments would further spread contamination; and the terminal facility would violate the U.S. Department of Housing and Urban Development (HUD) guidelines for acceptable separation distance from a hazardous facility.
>
> (*ibid.*, p. 4–184)

The DEIS reported that per capita personal income level for Baltimore County, Maryland and Chester County, Pennsylvania exceeds the level for the respective states. Turner Station, the residential community closest to the terminal site, is located 1.1–1.2 miles away. The median household income for Turner Station was reported as $28,324, which is greater than those reported for Baltimore County and the State of Maryland. Yet Turner Station is considered an environmental justice area, based on its minority population. Turner Station is part of the larger Dundalk community. Dundalk developed around past industrial activities such as steelmaking, shipbuilding, distilling, and others that have been declining. The report notes tax benefits and job opportunities, and positive cumulative economic benefits from the proposed facility. The economic analysis, however, is very limited.

The report concludes that the siting would not "disproportionately or otherwise result in adverse human health or environmental effects on minority or low-income communities or Native American programs. No identified minority or low-income population would be disproportionately impacted by the other projects considered in this cumulative analysis" (*ibid.*, p. 4-276). Suffice to say that there was considerable public concern expressed at meetings (see below).

## Conclusions about cumulative impacts

The DEIS identified multiple existing, approved, or proposed activities or projects in the region that could result in cumulative impacts. These overlap geographically to some extent. Seven are related to construction and operation of the LNG terminal, including a possible Sparrows Point power plant, an ethanol plant, a distribution facility, a highway widening, and upgrades of two wastewater treatment plants. Twelve are dredging projects with cumulative impacts on water quality of the Patapsco River. Another ten are associated with the proposed pipeline to be built by Mid-Atlantic Express Pipeline, and these include a wastewater treatment plant upgrade, an industrial facility expansion, five highway/road projects, a military base realignment/closure, and a natural gas pipeline expansion. Their DEIS assessment is that the impacts of LNG-related activities would be minimal and localized. They expect AES to use best management practices, and accordingly assume that the project's contribution to cumulative impacts on the waters crossed by both projects would be minor.

Regulatory authorities insist that oversight continue throughout the life of the project. For example, the FERC wants to make sure that at least 96 hours prior to arrival, each LNG ship would notify the terminal and the Coast Guard of its scheduled arrival. In addition, prior to entering Chesapeake Bay, the LNG ship would give advance notice to the US Navy, Patuxent River Naval Air Station, and other key parties. The language for operations and surveillance monitoring is similar to that for other notifications.

## DEIS conclusions

The major message from the DEIS, a typical one, is a conditional yes. The FERC does not expect an LNG release along the marine transit route, and if it were to occur, it would not significantly affect water quality in any of the surround- ing zones of concern because the LNG would quickly vaporize. An LNG release would temporarily and significantly impact water temperature to a limited depth under the created LNG pool. At the terminal site, no significant impacts are expected to occur among terrestrial or aquatic vegetation or along the proposed pipeline route. Yet, to minimize potential impacts to wildlife habi- tats, Mid-Atlantic Express is asked to prepare a management/mitigation plan.

After consulting with the US Fish and Wildlife Service and the National Marine Fisheries Service, FERC concludes that the proposed LNG project would have no effect, or is not likely to adversely affect threatened and endangered species or protected marine mammals, if their recommended mitigation programs are implemented. FERC also believes that essential fish habitat would not be significantly affected. It assumes that there will be impacts on aquatic organisms near the LNG facility from pressure waves associated with pile-driving activities during pier construction, from LNG boat traffic, and during water withdrawals for testing of LNG tanks and for ballast water for LNG tankers. FERC asserts that these impacts would be dealt with via agency-reviewed mitigation measures and would be short-term and/or minor. Normal operations of LNG tankers are not expected to have a significant impact on vegetation, wildlife, or threatened and endangered species.

Typically, draft EISs written by regulatory agencies end with conditional acceptance. This one clearly follows that pattern, containing conditions about public interest, construction, operation, mitigation measures, and the Cost Guard's safety and security measures:

> We have determined that if the project is found to be in the public interest and is constructed and operated in accordance with AES's and Mid-Atlantic Express's proposed mitigation, our recommended mitigation measures presented in section 5.2 of this draft EIS, and the Coast Guard's safety and security measures, construction and operation of the proposed facilities and the related LNG marine traffic would have limited adverse environmental impact and would be an environmentally acceptable action.
>
> (*ibid.*, p. ES-6)

The primary reasons for this conditional decision are also couched in similar, conditioned language:

> AES would construct its LNG terminal within an industrial port setting and the proposed pipeline facilities would follow existing, maintained rights-of-way for about 84.8% of the proposed pipeline route;

> AES and Mid-Atlantic Express would minimize impacts on soils, wetlands, and water bodies by implementing their ECPs;

> AES and Mid-Atlantic Express would be required to consult with federal and state agencies regarding the development of an ARMP that would compensate for impacts to wetland and water body resources;

> The Coast Guard's Waterway Suitability Report has preliminarily determined that the Chesapeake Bay can be made suitable for LNG marine traffic to the proposed facility, provided additional measures necessary to responsibly manage the maritime safety and security risks are put into place;

> AES would incorporate appropriate features and modifications, as specified by staff's recommendations, into the facility design to enhance the safety and operability of the proposed LNG facility;

The proposed facility would comply with the siting requirements of Title 49, CFR, Part 193;

AES would be required to develop and implement an Emergency Response Plan that would include involvement by state and local agencies and municipalities; include a Cost-Sharing Plan and a Transit Management Plan; and meet the requirements of the Commission, the Coast Guard, and other federal agencies;

AES and Mid-Atlantic Express would complete consultation with the SHPOs and the Advisory Council on Historic Preservation, as required by Section 106 of the National Historic Preservation Act, and with the Fish and Wildlife Service and National Marine Fisheries Service, as required by Section 7 of the Endangered Species Act, before beginning construction;

AES and Mid-Atlantic Express would obtain all federal permits and author-izations and would follow the applicable permitting requirements of the States of Maryland and Pennsylvania; and

The environmental inspection and mitigation monitoring program would ensure compliance with the mitigation measures that would become con-ditions to any authorizations of the proposed Project issued by the Commission.

*(ibid.,* pp. ES-6, ES-7)

## Reactions

In 2006, the FERC issued a Notice of Intent to Prepare an Environmental Impact Statement for the Proposed AES Sparrows Point LNG Terminal and Pipeline Project, Request for Comments on Environmental Issues, and Notice of Public Scoping Meetings. The notice was sent to approximately 2750 inter-ested parties, including federal, state, and local government officials; agency representatives; conservation organizations; local libraries and newspapers; and property owners within half a mile of the proposed LNG terminal and along the pipeline route.

FERC staff conducted two public site visits and held three open public EIS scoping meetings: one near the proposed LNG terminal and two along the proposed pipeline. They received over 500 comments from people, public officials, and government agencies. These comments expressed concern about public safety and security; siting options; dredging and disposal of dredged materials; impacts on fisheries, wildlife and vegetation; worries about boating and fishing disruption; wetlands and water body impacts; socioeconomic impacts; land use, residential and recreational impacts; air quality and noise impacts; and cumulative impacts. In essence, every one of the major impact categories in Table 4.1 was of concern to someone, but a great deal of concern focused on public health.

A few of the comments are extensive, but nearly all are brief. The essence of the struggle over the proposed LNG site is captured by articles reported in the local media. On April 3, 2006, FERC gave permission to AES and Mid-Atlantic Express LLCs to use the Commission's Pre-Filing Review Process for the proposed LNG facilities. The purpose of pre-filing is to provide an opportunity for FERC, the project sponsors, other federal, state and local agencies, and concerned citizens and nongovernmental organizations to identify and address project-related issues prior to the filing of an application with FERC.

On June 6, 2006, at a public meeting attended by the Governor of Maryland and senior county officials, hundreds of residents expressed concerns about the idea of living near an LNG terminal. The *Baltimore Sun*'s reporter (McMenamin 2006) noted that many of those testifying had lived near, and worked in, chemical, steel, and other facilities with industrial hazards, and yet this LNG facility was perceived in a harsher light. For example, the county fire chief reported that he did not think that it was possible to extinguish an LNG fire because it burns hotter than other fires.

AES attempted to present its proposal to the public and government bodies. On February 22, 2006, Kent Morton, AES's project director for the proposed Sparrows Point facility, and several colleagues briefed a Special Joint Meeting of the Sport Fishery Advisory Commission and Tidal Fishers Advisory Commission. The commissioners who spoke up were not favorably disposed toward the project. One stated that this project would allow AES to conduct its business while ruining the businesses of fishers (State of Maryland 2006).

On January 8, 2007, AES filed an application with FERC for authorization to site, construct, and operate an LNG receiving terminal and associated facilities. And on January 8, 2007, Mid-Atlantic Express filed an application for a certificate to construct, own, and operate an interstate natural gas transmission pipeline and ancillary facilities.

The response from opponents was immediate. Baltimore County attempted to exclude the location through its zoning powers. In January 2007, AES brought a federal law suit, asserting that the FERC had jurisdiction under the Energy Policy Act of 2005. In *AES Sparrows Point LNG, LLC v. Smith*, the court held that a change in the Baltimore County zoning ordinance prohibiting a new LNG facility in an area of the Chesapeake Bay that was part of Maryland's coastal zone management planning area was not pre-empted by the Energy Policy Act of 2005; that is, FERC did not have sole authority. The US Court of Appeals reversed the lower court's decision, granting authority to FERC. Notably, the court indicated that the zoning regulation had never been submitted to the National Oceanic and Atmospheric Administration for approval (Rosen 2007; Parfomak and Vann 2008). Governor O'Malley of Maryland asked the federal government to not allow the project (Wentworth 2007). Yet the US Supreme Court later supported the Appeals Court's decision. Next, the two Maryland US Senators introduced a bill that would strike down the provisions of the Energy Policy Act of 2005 that gave FERC the authority to pre-empt state and local powers to control energy facility siting (Schultz 2007). This effort was defeated.

AES did not draw back from the dispute. For example, it prepared responses to twenty questions posed by the Dundalk high school environmental science class (AES 2007). The almost seventy-page document, which was delivered in May 2007, answers each question and appends additional materials. The document emphasizes that Sparrows Point was chosen because it is the most remote site available, much more remote than the five existing US LNG sites. Since most students were concerned about LNG explosions, the document argues that LNG will not explode, and that fire is the worst-case event. Then it added that residences are too far away from the site for a fire to be a major concern.

On December 5, 2008, FERC (2008b) issued the Final Environmental Impact Statement (FEIS) with the Coast Guard, the US Army Corps of Engineers, and the EPA identified as key partners. It supported the proposed project and listed eleven reasons, each of which was noted above as part of the DEIS.

On January 13, 2009, Congressman Dutch Ruppersberger (Maryland, Second District) called on FERC "to delay decision on the proposed liquefied natural gas (LNG) facility at Sparrows Point until the Obama Administration takes office. . . ." Congressman Ruppersberger (2009, p. 1) clearly opposed the facility for this site and in densely populated areas, and decried what he considered a potential "rubber stamping [of] this ill-advised project."

On January 15, 2009, FERC approved the LNG terminal and pipeline proposal by a vote of four to one. Chairman Joseph Kelliher's words speak to the Commission's view of itself:

We have done so [approved the application] in this order, by attaching 169 conditions that will protect public safety and mitigate any adverse environmental impact and assure the AES Sparrows Point LNG Project will provide service in a safe and secure manner and provide fuel to generate electricity and heat homes. I realize this is not a popular decision, but it is the correct decision, rooted in a voluminous record and based on sound science.

(FERC 2009, p. 1)

Opponents responded that the decision is not final and would be tied up in courts for years, an assertion I have read many times, and the tactic has often worked (Ullmann 2009). Delay means increasing costs, and perhaps other LNG sites will be approved and eliminate the necessity for this site. Indeed, on September 22, 2009, the US Court of Appeals for the Fourth Circuit ruled that it upheld Maryland's decision to deny a water quality certification permit for the plant. Notably, the court focused on the dredging issue (see Ling 2010 for a summary).

# Interview

Lauren O'Donnell is Director of the Division of Gas – Environment and Engineering with FERC's Office of Energy Projects. Lauren has been with FERC since 1979, and she has been involved in, and in some cases responsible for, developing and implementing processes to carry out the Commission's responsibilities. For example, she helped design FERC's pre-filing process for both LNG facilities and interstate natural gas pipelines, and has managed the process since 2002. Given the regulatory role of FERC, I deemed it inappropriate for me to ask any questions about the specific project. I concentrated on generic issues during my conversation with her on August 20, 2009. It clearly was inappropriate for me to ask her about a politically contentious regulatory case that she was involved in. I had not met Lauren O'Donnell before this interview, and she was more than helpful in describing what FERC was being asked to do by the federal government.

My first question was: In light of criticisms of federal agencies for not coordinating with other government bodies in EIS preparation, how has FERC been able to coordinate the preparation of EISs that involve multiple federal agencies, state and local governments? Noting that FERC was assigned the lead in the preparation of EISs under the Energy Policy Act of 2005, Lauren O'Donnell focused on pre-filing and the establishment of EIS working groups. Both steps, she indicated, were taken to expedite receipt and analysis of applications. In 2002, the federal agencies involved with permitting natural gas facilities met to establish working relationships. Each agency independently has prepared implementation policies. She and her colleagues know their counterpoints in the Coast Guard, the EPA, DOT, and other federal agencies, so they do not have to waste time figuring out who has a role in the expertise of the cooperating agencies. What Lauren O'Donnell was doing for this EIS is precisely what some of the critiques described in this chapter called for.

The pre-filing process represents an opportunity to meet with other federal, state, and local agencies, applicants, and others to establish cooperation, and the parties are jointly able to flag the key issues. Beginning with the pre-filing, through scoping, the DEIS, and the FEIS, Lauren indicated that every effort is made by FERC to work with interested parties. Rarely, she noted, is there an issue among the federal agencies. However, it has happened that a federal agency, such as the US Fish and Wildlife Service, has not had staff resources to provide information for the EIS. In those instances, FERC will retain a consultant to do the work and then that information is provided to the cooperating parties for review.

States can be a bit more problematic, she indicated, primarily because sometimes they lack sufficient resources to participate fully. Even when state elected officials are against an application, she observed that state and local government departments almost always cooperate in providing information and critiquing documents. The clear message from Lauren O'Donnell is that

FERC will obtain the required information and secure as much input as possible from other agencies and parties.

My second question was about the role of applicants in providing information, and the question was based on the charge that the EIS depends too heavily on information from applicants, which is biased toward the proposed project. Lauren O'Donnell noted that applicants provide all the design data and other information required by FERC, and they pay for the services of consultants who provide information. Once the information is received, FERC staff scrutinize it, typically leading to rounds of requests for additional information. She emphasized that information and interpretations are not rubber-stamped.

With regard to the assertion that the EIS process causes unnecessary delays and costs a great deal of money, Lauren O'Donnell noted the process FERC uses was consciously designed to minimize delays. The pre-filing process can be completed in six months and may take up to a year and a half, depending upon the applicant's willingness to devote resources to the project. Within 90 days of a pre-filing, FERC will try to schedule activities so that all the parties understand what is expected, and when. A DEIS should be completed in six months, and an FEIS in another four months, with roughly another two months for comments.

She observed that some applications take less than a year and a half, but others take longer. Those proposed for the Gulf coast of the United States usually move faster because similar facilities already exist, pipelines cross timber or agricultural areas, and the public is familiar with infrastructure facilities and tends to view them as providing jobs and other economic benefits. In contrast, East and West coast applications typically take longer, she feels, because similar facilities are rare, pipelines may cross urban areas, and the public is unfamiliar with the facilities and concerned about safety and environmental impacts, and much less concerned about economic benefits.

Lauren O'Donnell views the FERC process as nearly optimized for efficiency. She noted that applicants did not complain about it because other EIS processes take much longer and require more resources.

My fourth question focused on how facilities could be assessed without the EIS process. She responded by indicating that other options could be gleaned from looking at legislative proposals for siting transmission lines. In essence, it is now recognized that the United States needs a nationwide capacity to transmit high voltage. A federal pre-emption-oriented group wants to increase FERC's transmission siting authority. State governments, however, are reluctant to grant any more authority to FERC. Over half the states have renewable portfolio standards, and the federal government may develop a renewable portfolio standard that will ultimately lead to more reliance on wind and solar power. In order to transmit the energy created by renewable energy systems, the country will need tens of thousands of miles of new transmission lines. States are extremely reluctant to grant FERC powers of pre-emption. Clearly, states have been reluctant to approve large-capacity transmission projects that cross state boundaries and in some cases even transmit energy from one part of their state to another. The result is delays and additional costs. When this chapter was initially written, there were four Congressional bills under consideration, and the controversy has continued.

Lauren O'Donnell's final comments were about the public's concern that it is not being heard. She observed that FERC's job is to gather and assess information as objectively as possible. Every effort is made to hear everyone, but, in the end, the Commissioners may respectfully reject the views of some participants. She noted that her group provides information to the Commissioners, and does not deal with the politics of the issues.

## Evaluation of the five questions

### Information

I found the presentation of some information in this EIS inadequate. As noted above, my biggest concern was with supply-and-demand issues regarding natural gas, and more specifically LNG (see also comments by FERC Commissioner Wellinghoff 2009, who expands upon these points). For context, over 30 years ago, I had the opportunity to examine EISs for a number of nuclear power plants proposed in the northeast United States. The proposed sites were urban sites, and there was widespread public and local government concern. Hired to advise the federal agency, I began looking at the supply and demand of energy estimates in the proposal. I found the estimates of demand to be much too high, and alternative supplies to be available.

In this case, I examined the EIA's *Annual Energy Outlook 2009* (EIA 2009). The EIA report would lead a reader to question the need for this LNG facility. With regard to domestic supplies, the EIA report suggests increases in LNG in the short run, but that there will be expanded use of domestic natural gas supplies, indeed that the United States will apply technology to extract more natural gas (itself a serious environmental issue). The report questions whether the United States can compete for imported LNG with other countries, for example, Korea and Japan. In essence, LNG imports depend on increasing prices in the United States. The reference case (intermediate demand) reports that LNG imports would rise to 6.5% in 2018 and fall to 3.5% in 2030 (*ibid.*, p. 42). Of course, there is disagreement among forecasts, which range from a 28% increase to a quadrupling of LNG imports.

On the demand side, I have to wonder about the potential impact of state renewable portfolio standards on LNG imports. From Massachusetts to Maryland, every state along the northeast coast has a renewable portfolio standard ranging from 15% to 24% (US Department of Energy 2009b). How these standards would impact the need for LNG is not clear, but there are no discussions of this possibility; rather, renewable resources are mentioned only briefly, and dismissed. I am perplexed by the failure to present a more thorough discussion. For example, other natural gas resources might come from Alaska, involving using oil shale or other technologies that might make this and other LNG facilities appear to be an environmental bargain.

Overall, the case can be made, as it is in the EIS, that more LNG facilities are needed, but the case can also be made that it is not justified by the supply-and-demand economic information presented in the document. The problem is a failure to address supply and demand in satisfactory depth. As noted earlier, opponents of these projects may struggle to understand the engineering of an LNG facility, but they will understand the supply-and-demand issues, as many of them are outspoken proponents of renewable energy portfolios. This failure to address these supply-and-demand issues in appropriate depth leaves the document vulnerable. Was this reluctance caused by the project-by-project focus of the Commission; by the lack of staff support; or other reasons that I can only guess at? My impression is that the Commission relies too heavily on the willingness of applicants to invest as a measure of supply and demand, rather than on its own capacity to pick and choose among sites.

My second concern is with the presentation of the safety analyses. I am not questioning the analyses *per se*, but rather the lack of information explaining the basis for the Coast Guard's requirements and regulations. Their presentation of numbers and maps without an explanation makes them appear arbitrary. The reader must trust the applicants and government to conclude that these discussions are correct. Unfortunately, some do not.

I am troubled by the mixing of science and political assertions in some statements. For example, the DEIS states:

Safety and security are important considerations in any Commission action. The attacks of September 11, 2001 have changed the way pipeline operators as well as regulators must consider terrorism, both in approving new projects and in operating existing facilities. However, the likelihood of future acts of terrorism or sabotage occurring at the proposed LNG import terminal, or at any of the myriad natural gas pipeline or energy facilities throughout the United States, is unpredictable given the disparate motives and abilities of terrorist groups. However, existing and proposed security measures discussed in this section make significant impacts to human life and property from a terrorist attack unlikely. The continuing need to construct facilities to support the future natural gas pipeline infrastructure is not diminished from the threat of any such unpredictable acts.

(FERC 2008a, p. 4-259a)

The final sentence must have had a chilling effect on some readers. The message is that we recognize there may be risk, but we need these plants, so we will need to live with risk. In this author's opinion, the sentence was not necessary and makes the siting decision seem predetermined. For those who separate the US population into income and political categories, that statement will be interpreted to imply that the needs of capital are more important than the needs of the local working classes, and more significantly, that FERC is not neutral in its assessment.

Third, I am not persuaded that there are locations to place the amount of dredged materials discussed in these EISs, and I remain concerned about public reaction to the transport of dredged materials in an urban area.

The tone and writing are accessible and the report is not too difficult to read; albeit the document is long and complex, and a great deal of patience is required to get through it.

## Comprehensiveness

The document includes a required list of health and safety, environmental, economic, historical, and social considerations, emphasizing the health and safety issues (Table 4.1). It also includes cumulative impacts. As noted above, I cannot tell how some conclusions were reached, based on what I read in this document. Some of the arguments are excellent: for example, the presentation of the technology of an LNG facility is sufficiently clear to be in a textbook. Other descriptions, as noted above, are inadequate.

## Coordination

This EIS is direct evidence that the federal government can mandate formal cooperation. The mechanism is to institutionalize it with memoranda of agreement between the agencies, and they will decide who will contribute what part to each EIS. Of course, this approach does not necessarily mean that the cooperating agencies all agree with the information and ultimate decision about the proposed action, but at least each agency had the opportunity to provide input and to critique the document. The pre-filing system enhances that opportunity throughout the life cycle of the EIS. States and local governments can be part of the process, but they have to engage in the process as cooperating parties, which some may choose not to do. Even if they choose not to engage with the cooperating parties, they are not precluded from participating.

## Accessibility to other stakeholders

All parties had access to the process, and all of their comments were publicly available on FERC's website. This is a major improvement in comparison with my past experiences with other EISs, in which documents were very difficult to secure and comments were not easy to submit (see Chapter 5). But were the hundreds of submissions meaningful? Access to process does not imply access to the decision. Part of the call for delaying the decision was that more time was needed to make meaningful comments about the document. Frankly, however, these documents are not that complicated. The call for delay is a call for an opportunity to bring political pressure to change the FERC's decision.

## Fate without an EIS

In the pre-1965 political climate of the United States, this project would have been approved. In today's risk-averse environment, left to the states, I do not think this project would have much of a chance of approval. Indeed, FERC's role in the process is clearly directed at overcoming state and local opposition. A speech on May 25, 2005 by J. Mark Robinson (2005, pp. 7–8), then Director of the Office of Energy Projects of FERC, is clear about the need to overcome local NIMBYism. I quote liberally from that speech, which was delivered before the Committee on Environment and Public Works of the United States Senate:

> Another issue of concern is the growing tendency for parochial, a local, interest trumping the greater public good. All siting is local and local concerns are of high significance, but if the standard for approving infrastructure requires that there be no local opposition for what in most instances are energy projects of regional importance, then no energy infrastructure will be built.
>
> *(ibid.)*

Mark Robinson then goes on to add landowners, towns, municipalities, and nongovernmental organizations to the list of potential opponents. He adds:

> Admittedly, much of the infrastructure proposed today is going to serve the future and those are comfortable with the status quo may not see any direct benefit for themselves. But if our parents and grandparents had taken that same attitude more than a half-century ago, I doubt we would be traveling on interstate road system we have today. We need a national natural gas system that contains a balance of domestic production and imported LNG deliveries, transportation, and storage. This system will serve the greater public at lower cost. There, of course, are legitimate local concerns, but to adhere to all of their requests to not be disturbed will result in a Balkanization of a national network it needs to expand and grow on an integrated basis.
>
> *(ibid., p. 8)*

Robinson's testimony used an LNG example in which six federal agencies, seven state agencies, and four local agencies had authority for a proposed LNG site. He then called for a "rational siting process" consisting of three elements: designation of one agency as the sole decision-maker; requirement that all agencies with authority over any element of the project work with the designated lead agency; and direct appeals of all actions at one time to the federal courts. This process is precisely the role that FERC now plays in the siting of LNG facilities. Is the EIS process illustrated in this chapter too bounded, and has it been captured by commercial interests? Is it an over-reaction to the criticisms about costs and delays attributed to the EIS process described in Chapter 1? Does it now preclude legitimate deliberation, or merely prevent deliberate stalling while waiting for political actions to undermine a legitimate proposal?

Readers of this case study are left with information about a specific technology and type of EIS. But perhaps we are also left with questions about how the EIS process has been adjusted to become an efficient planning process to meet a need that the administration of George W. Bush and the US Congress believed critical to the national economy.

# 5 Johnston Island: destruction of the US chemical weapons stockpile

## Introduction

Chemical weapons were first used during the First World War – with well over a million casualties and almost 100,000 deaths, including more than 50,000 Russians. The United States, concerned about potential use of these weapons against our troops, built a chemical weapons arsenal during the First World War, and that stockpile steadily increased. The United States began destroying the weapons during the late 1960s, under the so-called CHASE program ("cut holes and sink 'em"). In 1985, decades after building the second largest chemical weapons arsenal (the USSR had more weapons), Congress ordered the US Army to destroy its aging chemical weapons stockpile. Many of the weapons had become a threat to US soldiers, workers, and nearby populations. Congress was persuaded that some weapons could accidentally detonate and/or self-ignite. The 31,500 metric tons of weapons were stored at eight sites on the continental United States and at Johnston Island, an atoll about 800 miles from Hawaii (Figure 5.1).

The US stockpile consists primarily of vesicants (blistering agents) and organophosphorus nerve agents (Munro *et al.* 1999; National Research Council 2001). All of these chemical weapons can be lethal under certain conditions. Table 5.1 shows their major impacts on human health. The blistering agents may also be carcinogenic. Individuals who routinely handle these weapons told the author that they were more concerned about the vesicants than the others because of their blistering properties.

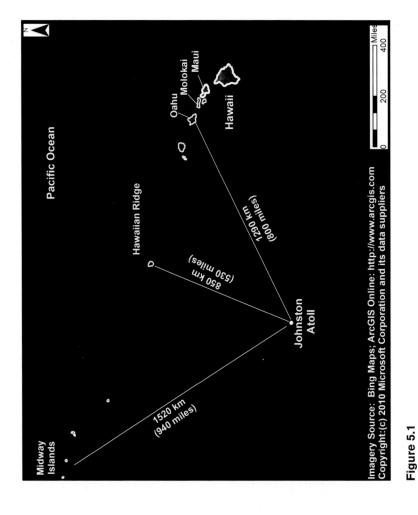

**Figure 5.1**

Location of Johnston Atoll in the central Pacific Ocean

**Table 5.1** Major chemical warfare agents

| Type | Agents* | Symptoms and health effects |
|---|---|---|
| Vesicant agents | HD (mustard)<br>HN (nitrogen mustard)<br>Lewisite<br>H (mustard with impurities) | Blisters develop in hours to days; eyes and lungs affected more rapidly; timing for vapors is 4–6 hours; skin effects at 2–48 hours |
| Nerve agents | GA (tabun)<br>GB (sarin)<br>GD (soman)<br>GF (cyclosarin)<br>VX | Potent cholinesterase inhibitor; difficulty breathing, sweating, drooling, convulsions, dimming of vision; incapacitates at low concentrations, kills at sufficient dosage; timing for vapors is seconds to minutes; skin effects at 2–18 hours |
| Respiratory agent | Phosgene | Difficulty breathing, tearing of eyes, damage to and flooding of lungs, suffocating, death; timing is immediate to 3 hours |
| Blood agents | AC (hydrogen cyanide)<br>CK (cyanogen chloride)<br>CN (sodium, potassium, calcium salts) | Rapid breathing, convulsions, and coma; kills at sufficient doses; nonpersistent, inhalation hazard, immediate effects |

*GA (tabun) = *N,N*-dimethyl phosphoroamidocyanidate; GB (sarin) = methylphosphonofluoridate isopropyl ester; GD (soman) = pinacolyl methyl phosphonofluoridate; H/HD (mustard) = bis-(2-chloroethyl)sulfide; HN (nitrogen mustard): HN1 = bis(2-chloroethyl)ethylamine; HN2 = bis(2-chloroethyl)methylamine; HN3 = tris(2-chloroethyl)amine; VX = *S*-(diisopropylaminoethyl) methylphosphonothiolate *O*-ethyl ester.

Number of weapons is a reason why some sites are more dangerous than others. The Tooele (Utah) site housed 13,600 metric tons, as compared with only 523 metric tons at the Lexington–Blue Grass (Kentucky) site. However, each ton is not equally hazardous (Table 5.2). About 60% of the original weapons stockpile had been stored in bulk containers (called "ton" containers), similar to tanks used for storing propane and other bulk liquids. These containers had no attached fuel or explosives, so they were less hazardous.

About 40% of the stored munitions consisted of more dangerous artillery projectiles, bombs, cartridges, land mines, mortar rounds, and spray tanks. They are more worrisome because fuel and explosives are part of these assembled weapons. The older weapons are troublesome. X-rays allowed us to see inside the weapon, and what we sometimes found was that the agent had decomposed into liquid, solid, and gas. The behavior of these compromised weapons, if disturbed, was less predictable than when the weapon was manufactured. Another issue is that some weapons were not perfectly manufactured and are more likely to leak. When found, these "leakers" are stored in another

**Table 5.2** Distribution of unitary chemical weapons stockpile, by storage location, February 2010

| Site | Tons | Agent* | Munition† | Disposal method | Percentage destroyed (Feb. 2010) |
|---|---|---|---|---|---|
| Tooele, UT | 13.6 | H/HD/HT, GB, VX | C, P, TC, R, B, M, ST | Incineration | 87 |
| Pine Bluff, AR | 3.9 | HT/HD, GB, VX | TC, R, M | Incineration | 64 |
| Umatilla, OR | 3.7 | HD, GB, VX | TC, P, R, B, M, ST | Incineration | 41 |
| Pueblo, CO | 2.6 | HT/HD | C, P | Hydrolysis/biotreatment | 0 |
| Anniston, AL | 2.3 | HT/HD, GB, VX | C, P, TC, R, M | Incineration | 67 |
| Johnston Atoll | 2.0 | HD, GB, VX | C, P, M, TC | Incineration | 100 |
| Aberdeen, MD | 1.6 | HD | TC | Hydrolysis | 100 |
| Newport, IN | 1.3 | VX | TC | Hydrolysis | 100 |
| Blue Grass, KY | 0.5 | HD, GB, VX | P, R | Hydrolysis/supercritical water oxidation | 0 |
| Total | 31.5 | – | – | – | 71 |

*See Table 5.1 for description of warfare agents; HT (mustard) = bis2(2-chloroethylthio)ethyl ether.

†B = bombs; C = cartridges, mortars; M = mines; P = projectiles; R = rockets; ST = spray tanks; TC = ton containers.

container (overpacked). Also, when the Army attempted to remove the explosives, the device that was supposed to unscrew the front-end did not always work because of manufacturing inconsistencies. The Army had to build a new device to remove the explosives, and meanwhile had to set aside these weapons. Much, but not all, of the entire aging stockpile was stored in earth-covered bunkers, colloquially called "igloos." The igloos could, under some conditions, be breached by an explosion.

Risk also depends on the fate and transport characteristics of the agents and their by-products. The Army developed rigid protocols to limit risk; for example, workers, air inside the destruction facility, outdoor air quality, stack exhaust, decontamination solution, fuels, use of protective suits, and equipment all must be tracked and managed (US Army 1988; National Research Council 2001). Chemical weapons destruction facilities have sophisticated monitoring equipment that can detect very low concentrations of agent. Workers are not permitted to enter rooms with agent unless they are in pairs, and they enter in sealed, impervious suits. The author was told of instances of employees being discharged from these high-paying jobs for what many would consider minor violations of these requirements. He witnessed one such incident, when a civilian contractor who could not find his gas mask borrowed another gas mask, and was fired for this violation of protocol. Nevertheless, no-one could say with 100% certainty that agent could not hurt the workers or escape the destruction chambers. Consequently, at Tooele, for example, carbon filtration was added to the emissions control system.

The number of people who could be exposed is another risk factor. Fortunately, nearly all of the weapons have been stored in remote locations. Johnston Island, the case study for this chapter, contains only military personnel and civilian contractors or visitors. Tooele, the site with the largest proportion of weapons, has some farms and homes within 10 miles (16 km) of the site, but the nearest large population cluster is Salt Lake City, more than 50 miles away. The riskiest situation involved Pine Bluff, Arkansas, where people lived within a mile of the site. Initial risk assessments at each site showed that doing nothing was riskier than moving and destroying the weapons. For example, estimates based on very conservative assumptions were one or more public fatalities in 20 years of continued storage at the Pine Bluff site at a probability of 3.03% (1 in 33), which compared with 1 in 20,000 in the case of on-site incineration of the weapons (US Army 1996, 1997).

While the main concern of the US Congress was human health risk associated with accidental or deliberate detonation of chemical weapons, international relations played a role in the call for destruction of these weapons of mass destruction. Following the devastating use of chemical weapons during the First World War, the United States signed the Geneva Protocol of 1925, and in 1972 it signed the chemical weapons treaty. Both signatures did not guarantee international compliance because neither addressed verification of weapon destruction. In January 1993, the United States signed a United Nations treaty, yet the US Senate did not ratify it until 1997. The delay was caused by members of Congress who asserted that some nations would not sign, and others would sign the treaty but violate it. Congress was also concerned about industrial espionage during inspections in the United States (General Assembly of the UN 1992; Committee on Foreign Relations 1996; Rotunda 1998; Yoo 1998). The UN treaty required that 1% of the weapons be destroyed by April 29, 2000. In fact, 15% of the US stockpile had been destroyed by 2000. In comparison, Russia, with a stockpile of 40,000 metric tons, did not meet the year-2000 deadline owing, it claimed, to a lack of funds (Greenberg 2003). By signing the international treaty, the United States agreed to eliminate the stockpile by April 2007, with the possibility of a 5-year extension.

Also, in 1990, the United States signed a treaty with the then Soviet Union that required 80% of their stockpiles to be destroyed, as well as production of chemical weapons to cease. Overall, the US Congress, a bilateral treaty with the USSR, and the UN treaty all required destruction of the US chemical weapons stockpile.

As part of its requirements, the US Congress required the Army to monitor progress toward destruction of the stockpile. The stockpile destruction program was to be overseen by the National Research Council, the US Department of Health and Human Services, the Centers for Disease Control and Prevention, the Environmental Protection Agency (EPA), the Council on Environmental Quality, the Occupational Safety and Health Administration, the Office of the Secretary of the Defense, and state and local government agencies. Furthermore, the Army required citizen advisory commissions to bring state and local public concerns about disposal to the Army, and promote public involvement at each of the eight continental US sites (National Research Council 2000a).

Several controversies have marked the program. The first was that, until the 1970s, the Army disposed of its unwanted munitions by burning them in pits. This caused mistrust on the part of communities. When the author visited some of the sites, residents reported feeling deceived and were reluctant to trust the Army. During the 1970s, the Army began considering neutralization and incineration. Incinerators were built on Johnston Island in 1990 (see below). Incineration has been criticized by some citizen groups (most notably the Chemical Weapons Working Group) and certain elected officials. The Chemical Weapons Working Group and others filed an environmental justice complaint with the EPA asserting that the Army was not fulfilling its mandate by opting for incineration. The Army did, in fact, use neutralization (e.g. hydrolysis) at several sites, but this also was not risk-free (National Research Council 1998, 2000b). Incineration was the main technology at five sites; hydrolysis and related methods were used at the other four sites (Table 5.2).

A second controversy at some of the sites was about emergency planning and management. At some sites, the Federal Emergency Management Agency (FEMA), which is responsible for the off-site plans, had a difficult time working with some state and local governments. For example, in 1996 the US General Accounting Office determined the readiness of the Anniston area for a chemical emergency event. The GAO concluded that the site was not prepared, that two-thirds of the money already allocated had not been spent, and that the Army, FEMA, and the state and local governments were in disagreement regarding fund amounts and allocation (GAO 1996). These problems continued, but were eventually solved and the weapons were destroyed (GAO 1997, 2001).

Cost has been a less controversial issue. In 1985, the cost of disposing of the chemical weapons was estimated at $1.7 billion (Lambright *et al.* 1998). Other estimates place the estimate at over $12 billion (WILPF undated), and this author's estimate is no less than $20 billion. In addition, there are large amounts of munitions and non-stockpile items that were part of the program, which will cost many more billions of dollars to destroy.

The destruction of the US chemical weapons stockpile represents a massive expense per dollar invested in the protection of the public's health (see also Chapter 6) for a high-consequence but very low-probability risk. The Army did not expect a series of technology decisions would be turned into a clash of values and morality, in which they would be portrayed as not doing everything possible to protect the public's health. The Army also has an ethical commitment to destroy these weapons because of the potentially stigmatizing effect of these stockpiles. In February 2010, the Army reported that 71% of the stockpile had been destroyed, including nearly all the fully armed and most dangerous weapon configurations, and including all the agent and weapons on Johnston Island (CMA 2010).

The remainder of this chapter focuses on the first place where US chemical weapons were destroyed: Johnston Island. I picked this location for three reasons. First, Johnston is about as remote a location as possible; therefore, it would be logical to assume there was little public interest. However, there was a substantial reaction to the Second Supplemental Environmental Impact Statement (SSEIS; US Army 1990a,b), which in ways described below was quite different from common NIMBY responses (see also Chapters 1 and 4). Second, Johnston was the first site to build and operate multiple incinerators for just these agents and munitions (and a variety of other supporting technologies). Johnston Island was an amazing engineering feat. Before describing the Johnston Island SSEIS, I should point out that I served on the Third NAS stockpile committee for about 6 years. However, I played no role in the Johnston EIS process described in this chapter.

## Johnston Island, the destruction of chemical weapons, and an ethical challenge

Johnston Island is an unincorporated US territory located in the central Pacific Ocean about 800 miles (1300 km) southwest of Honolulu, Hawaii (Figure 5.1). It is about 2 miles long and half a mile wide, and is about 6 feet above sea level. It is the largest of four islands: East, Johnston, North, and Sand (Figure 5.2). As

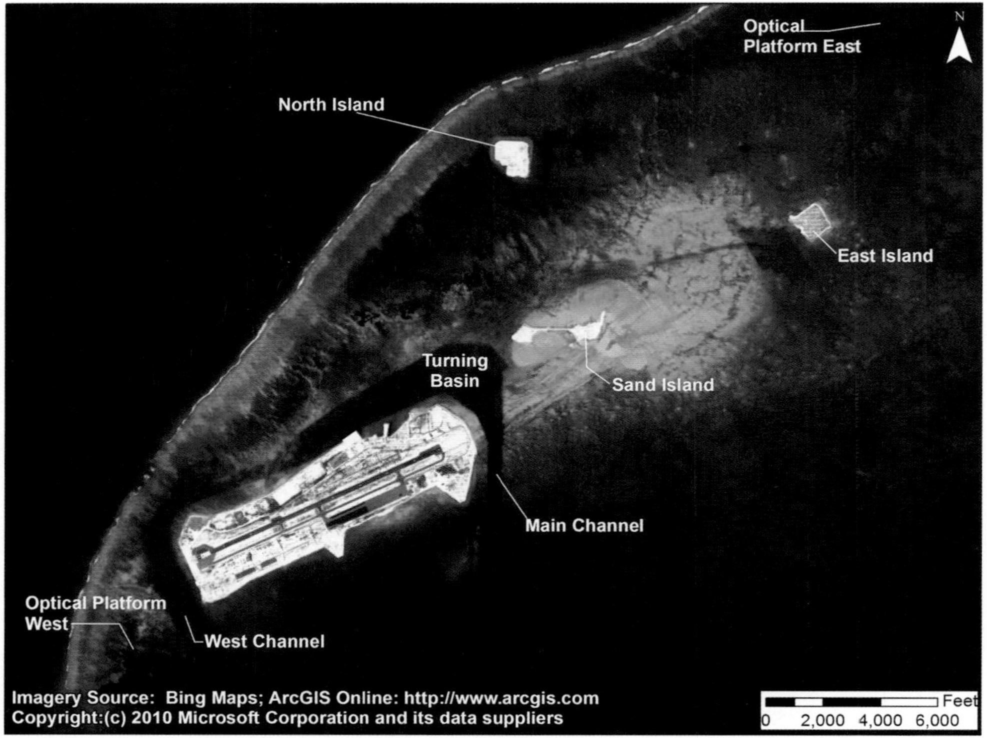
Optical
Platform East
N
North Island
East Island
Turning
Basin
Sand Island
Main Channel
Optical Platform
West
West Channel
Imagery Source: Bing Maps; ArcGIS Online: http://www.arcgis.com
Copyright:(c) 2010 Microsoft Corporation and its data suppliers
Feet
0 2,000 4,000 6,000

**Figure 5.2**

Johnston Island shipping channel and ship wharf area

a visitor to the island, the dominant visual impression is the runway that runs down the middle of the island. For the author, this was reinforced one morning while out jogging and faced with red warning lights flashing at the end of the runway, meaning get away from the runway because an airplane is approaching to land (Figure 5.3).

Johnston has an area where nuclear materials, non-chemical weapon materials and contamination are found; overall, the impression of the island is similar to being on the deck of an extremely large aircraft carrier. However, Johnston Atoll has been a National Wildlife Refuge since 1940. Several of the author's colleagues have asserted – and demonstrated – that the area surrounding the island has some of the best fishing immediately offshore in the Pacific.

Johnston Island already had chemical weapons before the SSEIS in this chapter was written: unitary chemical weapons moved from Okinawa in 1971. Unitary chemical agents and weapons are to be distinguished from binary chemical weapons. Binary weapons are built so that the toxic agent is not active in the weapon. The weapon has to be launched before the chemicals mix and become active, which makes it much safer before launch than the unitary munition. The unitary weapon contains the lethal or incapacitating mixture. Note that Johnston Island, even with its full complement of weapons, accounted for about 6% of the stockpile (see Table 5.2).

In 1983, the Army prepared an EIS for the construction and subsequent operation of an incinerator on the island (US Army 1983), and in 1988 it prepared the first SEIS (US Army 1988) to consider how to manage the wastes produced by the incineration process for the weapons already on the island.

In the case of Johnston Island, the life cycle of the destruction of chemical weapons moved to Johnston Island from Germany had six steps:

1   moving the weapons from existing storage near Clausen, in Rhineland-Palatinate, Germany, onto vehicles
2   moving the weapons to Bremerhaven, Germany for transfer to Johnston island
3   transport of the weapons on two ships to within 19.2-km (12-mile) territorial limits of Johnston Atoll, which took 1.5 months
4   unloading the weapons and bringing them to Johnston Island, where they were to be stored for incineration
5   incinerating the weapons
6   managing the waste products.

This SSEIS deals with steps 4 and 5, and more briefly discusses steps 1–3 and 6. By focusing on steps 4 and 5, as described below, the SSEIS arguably did not discuss the most serious risks, and the risks that most distressed public stakeholders.

The lead agency for its preparation was the Department of the Army (DOA), the Program Manager for Chemical Demilitarization. Five federal agencies cooperated in the preparation of the documents: the Defense Nuclear Agency;

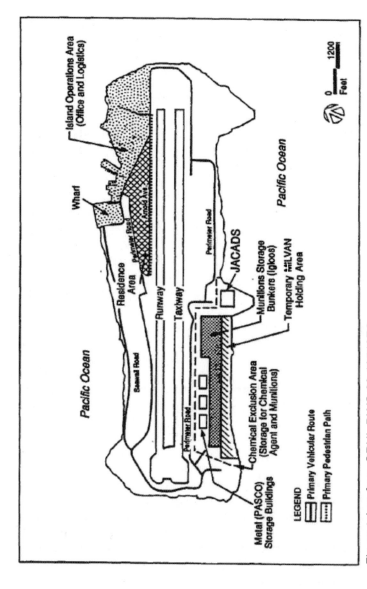

Figure taken from ORNL-DWG 89-19513R2

**Figure 5.3**

Location of facilities and major roadways on Johnston Island

the Department of Commerce, National Marine Fisheries Service (NMFS); the Department of Energy (DOE); the Department of the Interior, Fish and Wildlife Service (FWS); and the Environmental Protection Agency.

As mentioned above, the ethical dilemmas inherent in this proposed action and underscored by this SSEIS were the major reason for choosing this case study. I raise two questions at this point in order to underscore the issues. Was it appropriate to include only minimal information about the first stages of the life cycle of the destruction of the agents and munitions? With regard to this question, not every EIS in this volume covers the entire life cycle of an action. For example, Chapter 4 examines the construction and operation of an LNG facility near Baltimore, Maryland. That EIS did not include the impact of securing the natural gas from the ground (risk and impact outside the United States), nor the risk of loading it on a massive ship. It did not discuss risks associated with the ocean voyage. The chemical weapons SSEIS has the same omissions. The chemical weapons SSEIS says that:

> other phases of the movement of the European stockpile will be addressed in separate environmental documentation. Environmental analyses of the movement from existing storage to port in FRG is the responsibility of the host nation. The ocean transportation to the 19-km (12-mile) limit of Johnston Atoll and possible alternatives have been assessed in a separate Global Commons environmental assessment under Executive Order 12114.
> (US Army 1990a, p. 23)

A stated objective of the Johnston Island chemical weapons project was to destroy the Asian chemical munitions already on the atoll. But the second piece of that technological challenge was that it was to serve as a proof-of-method demonstration of the Army's preferred method of disposing of the stockpile of unitary chemical agents and munitions. Is the blatant statement and implementation of that principle in this location appropriate, given the history of the area?

## Preferred action

The preferred action was to incinerate the stockpile of NATO chemical munitions at Johnston Island. The SSEIS reported that these munitions consisted of about 100,000 munitions of 8-inch and 155-mm projectiles containing VX and GB nerve gas agents. These weighed approximately 430 tons. As noted earlier, the proposed action does not begin with moving the weapons from their locations to Johnston Island.

Nevertheless, beginning at the start of the process, the Army placed the munitions on pallets, and these in turn were packed in containers. The inner of two containers is built to be vapor-tight and meets International Maritime Dangerous Goods Code requirements. In turn, these were packaged in military vehicles (so-called MILVANs) that are not airtight for further control and protection. Two ships accompanied by US naval vessels (route kept secret) brought the ordinance to within 12 miles of Johnston Island.

The SSEIS officially begins with the two ships being guided by tug boats to the atoll's wharf. The naval escort stays with the second ship while the first is unloaded. The process of transferring the munitions to the island, the report says, takes 2 weeks.

The second phase is to unload the two ships containing the weapons. Each ship's on-board overhead cranes, the report says, are used to offload the MILVANs to transport trailers. An important step in this process is that each MILVAN is monitored for leaking agent. Any found to be leaking are moved immediately to the chemical exclusion area on the atoll and decontaminated. The MILVANs are then to be transported to the chemical exclusion area while guarded by military police.

The third phase is placement of the weapons in the chemical storage area until they are incinerated. The area is heavily guarded and monitored by lighting, and protected by barricades and intruder detection devices. Some munitions already on site will be destroyed first, and then the European munitions will be placed in the on-site igloos. The report estimates this will take 120 days. The note expects the actual incineration will take 3.5 months.

The last phase requires removal of the MILVANs (they are reused) and movement of waste to a site on the continental United States for disposal.

The SSEIS focuses on five impacts, and judges all of them to be insignificant:

- increase in land use for temporary holding of MILVANs
- increase in the total number of chemical munitions in storage
- increase in the total emissions and wastes from 3.5 additional months of operation
- temporary increase in the number of workers residing on Johnston Island
- increased potential for an accident that could release chemical agent into the environment.

Because the document has no major impacts to report, it makes no sense to go through the exercise of reviewing every potential impact. Table 5.3 lists the major categories of impact considered in the SSEIS.

It does make sense to illustrate the carefully crafted responses in the document. For example, here is a verbatim presentation on threats to endangered species:

The only rare species of reptiles in the Johnston Atoll area are sea turtles. The threatened green sea turtle (*Chelonia mydas*) and possibly the endangered hawksbill turtle (*Eretmochelys imbricata*) have been observed feeding at Johnston Atoll, although nesting is not known to occur. Green sea turtles are much more common and are often observed feeding along the south shore of Johnston Island. It is estimated that 200 green sea turtles use this

**Table 5.3** Environmental impacts considered for Johnston Island EIS

| | |
|---|---|
| Water resources | Groundwater |
| | Ocean water |
| Air quality | |
| Ecology | Terrestrial |
| | Aquatic (coolant water, stack emissions) |
| Waste management | Liquid process waste |
| | Solid waste with value |
| | Solid hazardous waste with no value |
| | Other wastes |
| Analysis of potential accidents | Agent releases to the atmosphere (effects on humans and ecology) |
| | Spilled agents (effects on humans and ecology) |
| Worker safety and agent monitoring land | Occupational safety and health |
| | Monitoring (agent exposure limits, instrumentation, proposed agent monitoring and response to agent incidents) |
| Cumulative impacts | Air quality |
| | Land use |
| | Groundwater |
| | Waste production and disposal |
| | Terrestrial ecology |
| | Aquatic ecology |
| | Pipeline safety standard |
| | Pipeline accident data |
| | Impact on public safety |
| Unavoidable adverse environmental impacts | |
| Irreversible and irretrievable commitment of resources | |
| Short-term use and long-term productivity | |

area of Johnston Atoll as a feeding ground. NMFS studies done in 1983, 1985, and 1987 found that green sea turtles at Johnston Atoll were more healthy and robust than turtles studied in the northwest Hawaiian Islands. Three turtles tagged in that study were subsequently sighted in the Northwest Hawaiian Islands. From these sightings. -. . . The Johnston Atoll and Hawaiian Island green sea turtle breeding population are the same. The resident FWS managers state that the turtle feeding ground near the sewage outfall has one of the highest concentrations of green sea turtles at DOD-nesting foraging grounds in the Pacific.

Unlike other islands in the Pacific where the green sea turtle is protected but still taken for food by local inhabitants, the combination of strict government control of Johnston Island and resident FWS personnel ensures protection of the local green sea turtle population.

*(ibid.,* p. 55)

The impression is that the Army is protecting the species, which indeed is likely to be the case.

Not surprisingly, on-site personnel were concerned about occupational exposures and safety. Here is a flavor of what the document says about this issue and what already happened on it:

Numerous health and safety plans, including the JACADS [Johnston Atoll Chemical Agent Disposal System] Safety and Occupational Health Plan and the Johnston Atoll Medical Support Plan, establish procedures to ensure the health and safety of individuals working in munitions disposal operations. All JACADS personnel are given training and information on hazardous chemicals in their work area before their initial assignment and whenever a new hazard is introduced into their work area. On-going training topics include accident prevention; wearing, adjusting and caring for protective masks and clothing; emergency procedures; and decontamination procedures. All personnel arriving on Johnston Island are briefed on chemical agent effects; they are issued and fitted with a protective mask.

Agent monitoring . . . is provided to ensure rapid response if leaking agent is detected, in which case procedures from the CAIRA Plan would be initiated. These procedures define activities to protect workers and to respond to a chemical release. The Army has conducted a preliminary assessment of potentially hazardous areas on Johnston Atoll to determine the nature and degree of threat posed by these areas and to identify areas that may require immediate cleanup activities. Two areas of concern are a former herbicide ("agent orange") storage area and an area containing plutonium-contaminated soil. The former herbicide storage area is located on a peninsula at the northwest side of the island. Previously, herbicide was inadvertently spilled in this area. Access to this area has been restricted by fencing. All island personnel are briefed about the restricted area and Johnston Island security personnel routinely monitor the area. The US Air Force has responsibility for cleanup activities in this area. The plutonium contamination resulted from the malfunctioning of missiles during the 1962 atmospheric nuclear tests conducted by the United States.

*(ibid.,* pp. 73–74)

An interesting footnote to the above description is that David Gibbs, a maintenance pipe welder, was killed on November 27, 1997 when a piece of equipment fell on him that was part of the Deactivation Furnace System. The accident analysis showed that the contractor had not been instructed in the proper safety practices for this task. The author knows of no other fatalities at Johnston.

## Accidents

Prior to the SSEIS, the US Army (1983, 1988) had analyzed 3000 possible accidents, rating each by likelihood, the potential amount of agent that could be released, the type of agent, how it could be released (spill, detonation, or fire), accident locations, and the duration of time of the release. The SSEIS lists one "worst credible accident" for:

- storage in igloos
- disposal and handling operations
- interim and continued storage
- MILVAN handling and transport.

The EISs concluded that the accidents have likelihoods of less than one chance in a million. If they occurred, then there would be some potential health effects. The current SSEIS presented a more sophisticated modeling analysis than did the prior version, and this presentation was enhanced by the presence at the public hearing of an expert from the Oak Ridge National Laboratories, who carried out plume simulation models.

The report presented two types of accident analysis: one is a detonation, fire, and/or spill leading to an airborne release; the second is a water emission resulting from a spill. For example, the report considers a plane crash into a metal building holding agent, leading to a spill, as the most serious water accident. The report acknowledges the possibility of serious damage near the spill, but states that the agent would dilute in the open ocean. The modelers conclude that, in the case of an airborne emission, lethal exposures could extend as far as approximately 62–76 miles for the air and 14 miles for the water emission, but would not reach Hawaii (these were worst-case scenarios).

In this regard, the report notes:

The accident analysis presented in this section involves many conservative (i.e., pessimistic) assumptions. Worst case accident scenarios and worst case meteorological conditions are used to establish the absolute upper bound on any potential effect of chemical agent accidentally released to the environment. Those accidents are extraordinary events with an extremely low likelihood of occurrence; it is further assumed that no actions are taken to control or mitigate the consequences of such an accident. The result of these assumptions is the identification of a hypothetical zone of potential impacts that can be used to describe the boundary beyond which adverse environmental impacts would not be expected to occur.

(*ibid.*, p. 63)

Activities and land uses on the crowded atoll were planned at least partly to protect the 1200+ occupants. Unloading from the wharf would not occur when winds could injure personnel. The storage areas were located downwind from housing. With regard to the environment, the SSEIS reports that the bird population nests upwind but feeds downwind of the incinerator site. A release could harm them, depending on the amount and time of exposure. The endangered humpback whale and Hawaiian monk seal are noted to visit this area, but the report concludes that these visit are so infrequent that the risk is low. Endangered green sea turtles are found in the area, and could be injured by a spill or atmospheric emissions. The NMFS and the FWS were reported to be advising the Army as part of their cooperating role status regarding ecological issues.

## Cumulative impacts

Cumulative impacts can result from individually minor but collectively significant actions occurring over a period of time. Review of past activities and reasonably foreseeable future activities at Johnston Island allowed for a qualitative cumulative impact assessment in the following six areas:

- Air quality: destruction of the European stockpile means 3.5 months more of operations. The emissions will remain the same per unit of time, but increase 20%.
- Land use: 1.2 hectares (3 acres) of land will be needed for the temporary holding of munition-filled MILVANs in the exclusion area, and perhaps 1.2 hectares (3 acres) for the temporary holding of empty MILVANs. This is far less than 1% of land on the island. The report says that the land will be returned to its previous state after the campaign is completed.
- Groundwater: Johnston Island has no potable groundwater.
- Waste production and disposal: all the waste is to be removed from Johnston Island and disposed of at an approved site on the continental USA.
- Terrestrial ecology: there will be temporary loss as a result of MILVAN holding and road surfacing. The report indicates that there is little seabird activity in that area.
- Aquatic ecology: no adverse impacts are expected, and construction of a new waste treatment plant is expected to have a positive effect on water quality.

The conclusion is that there are no significant cumulative impacts on Johnston Island resources. No unavoidable adverse environmental impacts were expected.

Labor, materials, and capital spent on the preferred action were described as an irretrievable commitment of resources, but seemingly given little weight.

## Alternative actions

More than any case study in this book, the alternatives were critical here, and yet were essentially dismissed. The no-action alternative was to leave the weapons in Germany in their storage areas. That option was rejected because in 1986 President Ronald Reagan agreed with Chancellor Kohl to remove them from Germany by 1992. In 1992, Secretary of State Baker and Chancellor Kohl agreed to speed up the removal to no later than December 1990. In essence, the Army had to move the weapons from Germany, so there was no German-based option.

Legal constraints severely limited options that had not already been precluded by political decisions. The SSEIS interprets the Federal Public Law 99-145 as favoring rapid destruction of the munitions. In fact, the schedule was not followed in many continental US locations. Transfer of the European chemical munitions was prohibited until the Secretary of Defense certified to Congress that there was sufficient storage capacity for the European weapons and that the Johnston Atoll disposal system had destroyed live agent chemical munitions. Once the stockpile left Europe, alternatives that may have been considered included disposal at facilities other than on Johnston Atoll, and interim storage at locations other than Johnston Island. In addition, a number of options within these major alternatives could have been considered. These included alternate transport modes (e.g. air or sea), alternate routing, and alternate disposal technologies. But they were not considered in detail in the SSEIS.

The report also briefly discussed, and then dismissed, the eight continental United States sites as inappropriate, given the prohibition against bringing the weapons to the United States and even transferring them among the eight sites. The report argues that each site will have technology that fits its weapons. (See above for discussion of sites and program cost.)

The most compelling reason for not using the US facilities was the prohibition on the movement of chemical stocks into the country from outside the continental United States. The document dismisses other European locations as unsuitable because there is no place to store the weapons, much less destroy them.

The report concludes that only the Johnston Island site can most effectively handle the European stockpile. The Johnston Atoll site is asserted to be the best currently available and readily accessible site for destroying the agents and weapons. Entirely new storage and disposal facilities could not be constructed at non-US locations without building new facilities and getting approvals, which doubtless would have proven difficult, if not impractical or unfeasible. US-based destruction was precluded by the requirement that the technology was to be demonstrated on Johnston Island before being used on the continental United States. In reality, the political process apparently left no politically and legally actionable options.

Given the restricted scope of the SSEIS analysis, which means assuming that the agents and weapons were anchored off the coast of Johnston Island, the impacts of adding the additional ordinance to what was already on the island leads to the rational conclusion that the impacts, as interpreted by the SSEIS, are relatively minor. For example with regard to construction, no additional buildings are proposed. The document suggests that some road and soil stabilization will be required in the chemical area, which will also mean some additional dust generation. In contrast, any other option, other than keeping the ordinance stored in Germany at its current location, would have required building new facilities.

Incineration is preceded by disassembling the ordinance, that is, removing the explosives from the assembly, draining the gas, and removing the fuel. The facility has four different incinerators to handle this process and a feed system that was specifically designed for these purposes. There is a liquid incinerator for the gas, a deactivation furnace for the explosives and fuel assembly, a metal parts furnace for metals that have been in contact with the gas, and a dunnage to burn the other wastes. Each one has a pollution abatement system. These specially built systems were to be tested on Johnston Island before being adapted for other locations. The marginal impact and risk of destroying the European weapons is minimal compared with what it would be for using any other location that did not have these technologies already in place. The document notes that the predicted concentrations of contaminants will be well below National Ambient Air Quality Standards (NAAQS) for criteria pollutants, and meet the rigid high standard for destruction of chemical agents. In comparison, the other alternatives would have required building systems elsewhere (four of the continental US sites did build incinerators), or would have had to wait for another technology to be developed and then tested (four did), meanwhile assuming that none of the aging ordinance would detonate.

The Army planned to reuse the MILVANs and decontaminated scrap metal and to package and ship solid wastes to the continental United States for disposal. At the time of the hearing, the Army was negotiating to find the best option. Hence the public did not know where the waste products were to be shipped. In short, none of the alternatives was given serious consideration in the document; by precluding other major alternatives, primarily for political reasons, they were left with Johnston Island.


## Compliance with regulations

The document took great care to demonstrate that the Army had been complying with federal legal requirements. The SSEIS notes that it initiated informal scoping meetings on the Johnston Island project in February 1982 with the EPA, FWS, and NMFS. More meetings were held in May and July 1982. In March 1983, scoping workshops were held for federal, state, and local agencies and the general public. In February 1983, the Army issued a Notice of Intent to prepare an EIS for the project, which was published in the *Federal Register*, and the Army issued formal press release. In July 1983, the Draft JACADS EIS was distributed and the Notice of Availability for that document was published in the *Federal Register*. The Final EIS was published in November 1983, followed by a Record of Decision in December 1983.

In 1983, the Army began considering the option of ocean disposal of the waste generated by the Johnston Atoll facilities. In November 1985, a public hearing was held in Honolulu, Hawaii, to elicit public input on the proposal for ocean disposal. Then the Army decided to prepare an SEIS to evaluate disposal alternatives for the process wastes. A Notice of Intent was published in the *Federal Register* on April 30, 1987. The draft SEIS was distributed in October 1987. Comments were received, and the final version was published in November 1989. A Record of Decision was issued on December 27, 1989. The notice to prepare the second supplemental (SSEIS) to deal with the wastes from Germany was issued in September 1989. In February 1990, the draft SSEIS was distributed. The public meeting was held in Honolulu on March 20, 1990, and is summarized below. The report very carefully, and I am sure quite purposefully, listed and described more than two dozen federal laws, regulations, and executive orders with which it had complied.

## Public response

The Johnston Atoll EIS attracted about 100 people to a public meeting in Honolulu, Hawaii on March 20, 1990. The text was 519 pages, of which twenty-seven came from nine federal agencies and the remaining 492 from more than two dozen spokespersons and individuals. The federal respondents had been cooperating agencies. For example, the NMFS was satisfied with the Army's response to its earlier suggestions, and offered a few more. The EPA asked for more information, which is typical. They offered ten comments or questions asking the Army to explain the legal reasons why the European stockpile had to be moved, emphasizing the need for as little interim storage as possible, suggesting that their Resource Conservation and Recovery Act permits would need to be changed, and asking questions on worker exposure and safety.

Jonathan P. Deason, US Department of the Interior, Office of Environmental Affairs noted that the transportation, storage, and disposal would have little direct impact, and yet he noted some concerns about potential impacts from human waste disposal associated with additional workers. Representatives of the DOE, the Defense Nuclear Agency, and the US Public Health Service had little to add.

In contrast, the public representatives had a great deal to submit to the record, and none of it was supportive of burning the weapons on Johnston Island. Rather than review each one, I have divided the comments into four themes in order of the frequency of their utterance.

## Availability of documents

The most often asserted argument was that the Army had not provided the documents for review in sufficient time for public review. Several people testified that they obtained a copy only a week before the public hearing. One testified that the Governor's Office of Hawaii and the Health Department received copies of the document from her, not from the Army (US Army 1990b, p. 445). Her statement, while more aggressive than others, is consistent with the sentiment as judged by the transcripts. Marsha Joyner testified:

> This whole secrecy has been too much. We published the notices, we called the newspaper, we handed out the EIS, and for anybody that got one, it came from us. I think since the Army did such a piss-poor job, that we are entitled to at least a 90-day period when everyone in the State, especially the Health Department, the Civil Defense, the Governor's Office has the time to read this. And to go any further is ludicrous, and, as everyone has said. If it does go any further, it means that it's a done deal and you have lied to us again.
>
> *(ibid.,* pp. 445–6)

A number of respondents asserted that the failure to make the document available long in advance of the meeting was a deliberate effort to circumvent the legal requirement. No-one representing the Army debated the charge; they could not explain why, for example, it was not available in public libraries, and they promised that they would look into what had happened. Some of the 100 attendants concluded that the Army would not even make the transcript available, despite its spokespersons indicating that it would be made available. This incident was embarrassing for the Army.

## Environmental injustice

A second theme was that the transfer of weapons from Germany to Johnston Island was an environmental injustice. Marsha Joyner, a Hawaii resident, made a statement that requires no interpretation:

> the fact that you can move this convoy of 400 tons of lethal nerve gas across thousands of miles of ocean and around non-white nations and not notify anyone amounts to environmental terrorism. This is racism at its highest level. The thing about racism is that it is so insidious that you don't even know that you are doing it. You don't see us. The military record of people treatment in the Pacific since martial law was declared has been one of shame. I find it hard to believe that this is any different. In preparing the EIS the statement was made that it (Kalama Island) is away from population centers. What you neglected to say is white population centers.
>
> *(ibid.,* p. 82)

Others added that they felt like guinea pigs. Several noted that a spill could jeopardize the perception of the growing tourist industry in the region and undermine the region's growing economy.

## Incomplete science

The third theme focused on incomplete science. Presenters argued that movement of the waste was irrational. They noted that, if the waste was so safe, why not incinerate it in West Germany, where it originated. If it is so safe, why is its path across the oceans a secret? Greenpeace presented a 100+ page report, arguing that it wanted the waste stored in Germany until a safer disposal method was available. Many of the attendees were persuaded by the Greenpeace report and referred to it in their comments. Hawaii legislator Annelle C. Amaral stated:

> The conclusion is clear: the risks created by transporting the chemical weapons from Europe to Johnston island far outweigh the marginal benefits of disposing of these weapons at Johnston Atoll versus disposing of them on-site in Europe or waiting for better solutions in the future.
>
> (*ibid.*, p. 29)

Various presenters questioned specific elements of the science and engineering. For example, a meteorologist questioned the plume models used to simulate contaminant dispersion, arguing that the models lacked sufficient information about local weather conditions, and that in fact concentration of lethal agents over the Hawaiian Islands could be higher than estimated in the Army's computer models. Several commentators argued that Johnston Island had to be evacuated twice because of hurricanes (Celeste in 1972 and Keli in 1984), and they wanted Army to explain what it would do if the island and igloos containing the weapons were swamped by a hurricane; what would happen if a hurricane came while ships carrying the chemical weapons were waiting to unload their hazardous cargo; and what the Army would do if there was a spill.

Other comments focused on the potential impact of the action on surrounding waters, and on sea turtles, fish, whales, and the atoll's bird sanctuary. Two speakers were worried about the transport of water brines from the site to the continental United States. Several asserted that the German government lacked experience in moving these kinds of weapons, and still others were worried about the stability of the weapons and the effort to manage them on the island before incineration. Several speakers referred to nuclear hot spots from an accident in 1962 with a nuclear-tipped warhead firing, and Agent Orange hot spots on the island (a 1977 spill on a ship that was incinerating Agent Orange offshore), and were concerned about protecting those on the island.

## Legality

The fourth theme was that the EIS was violating the law. Two attorneys testi-fied and challenged the legality of the action. Paul Spaulding, III, representing the Sierra Club Legal Defense Fund, submitted a fourteen-page memo arguing that the Army had violated NEPA by not addressing the movement of agents from Europe to Hawaii. In essence, he argued that the EIS was not a compre-hensive life-cycle analysis of the waste management issues because it focused on destroying the gas on the island, while ignoring the transport from its location in Germany to the atoll. He asserted that the real question is, how should the waste currently in Germany be destroyed?

Attorney Spaulding continued that the US Army had bowed to politics, that in 1986 President Reagan had promised Chancellor Helmut Kohl of Germany that the United States would move its entire European (NATO) chemical weapons stockpile out of Germany by 1992, and earlier if possible. Spaulding asserted that Chancellor Kohl's desire to move the weapons out of Germany was insufficient reason for the US Army to move them to Johnston Island.

He wanted more attention focused on the no-action alternative – leaving the weapons in Germany until a better disposal method could be developed, or destroying them in Germany or elsewhere at a NATO location. He pointed out what he considered an inconsistency by arguing that the EIS rejected the Aberdeen Proving Ground in Maryland as a site because it is near Chesapeake Bay, and for various other reasons. And yet, he wondered, how are these same concerns not present at Johnston Island? If it is risky, he argued, to transport the weapons to Maryland, then how could it not be risky to transport them to Johnston Island? If the species in Chesapeake Bay are at risk, then are not those surrounding the atoll and on the atoll's bird sanctuary equally at risk? The preferred solution, he argued, is not legal. He added that "Whereas the public ordered the environmental equivalent of a 'filet mignon' steak, the Army has now served it with a 'ground beef'" (*ibid.*, p. 34).

It did not escape the two attorneys that whereas nerve gas on the conti-nental United States has to be destroyed on site, that is, no transport is allowed across state boundaries, it is permissible to move it halfway around the globe from Europe to Johnston Island. One attorney, Jon M. Van Dyke, argued that he was not persuaded that this action was even legal. His argument was that the two original islands were only a little more than 50 acres when the United States seized the atoll from the Kingdom of Hawaii in the nineteenth century; through dredging and dumping, the island had become almost 700 acres in size. Van Dyke asserted that the atoll, to all intents and purposes, is a man-made structure at sea, and the 1972 London Dumping Convention, ratified by the United States, does not permit the destruction of waste on bodies at sea, including platforms or man-made structures. Many of these arguments were thoughtful and well put.

# Interview

David Kosson is Professor and Chairman of Civil and Environmental Engineering at Vanderbilt University. His specific research has focused on mass transfer of contaminants in soils, sediments, and wastes; development of innovative remediation processes for waste sites; methods to test for leaching; and many others. More broadly, Professor Kosson has been a leader in national efforts to manage nuclear waste and chemical weapons. Apropos this chapter, he was a member of the National Academy of Science's chemical weapons so-called "stockpile committee" for two full three-year terms, and he chaired it during a period when the committee visited sites not only to examine the technical issues, but also to listen to public viewpoints. I would find it hard to think of anyone who has a better understanding of the technology and the public–political side of this issue. I interviewed him on February 24, 2010.

I asked Professor Kosson to assess the complexity of the engineering tasks at Johnston Island. He responded that each of the four incinerators operated relatively independently. The greatest risk was dismantling the weapons and feeding them into the system (see above for a description). The facilities are built so that the most dangerous areas are in the center. The hot zone is heavily monitored and secured. Then there is a buffer zone, and finally an outer zone. The intensity of monitoring decreases with distance from the inner hot zone.

Professor Kosson believes that the technology built on Johnston Island could have been built without demonstration on Johnston. However, it was helpful to have had the demonstration first on the atoll because lessons were learned about systems operations, future design improvements, instrumentation, and about handling the ordinance, agent, and waste on Johnston. If it had been built first at a continental United States location, he thinks that they might have first needed to build a demonstration project, rather than a full-scale operating plant. Professor Kosson praised the Army for "groundbreaking efforts" to use probabilistic risk assessment and linking the risk assessment to its risk management practices. Overall, based on his public experiences at the continental United States sites, he feels that it was sensible to have a demonstration in a less populated area in order to gain both technical and public confidence.

With regard to the risk of removing the waste from Germany and shipping it across the globe, Dr Kosson believes that the US Army treated this shipment as it would the movement of any weapons, which means that it was handled by weapons experts, packed by experts, and guarded by military personnel. While handling weapons is always a risk, he reiterated that the greatest risk was disassembling the weapons and the movements associated with those processes – as well as doing nothing, that is, allowing the weapons to age, and potentially leak or have accidental auto-ignition.

To emphasize the relative significance of risks, Professor Kosson noted that, in light of the successful destruction of the weapons at Johnston Island, Tooele, and other incineration sites, the Army had proposed using incineration at other sites. However, some local communities had opposed incineration. For example, the Blue Grass, Kentucky site (see Table 5.2) rejected incineration in favor of a set of technologies that had not been demonstrated. Consequently, the destruction of agents at that site, stored primarily as fully configured rockets, will be delayed a decade. Meanwhile, the weapons are aging in place, which poses a greater risk than using incineration or another proven technology.

## Evaluation of the five questions

### Information

The focus of this document is a proposed supplementary action. A supplementary EIS should not be expected to be as thorough about the entire project as the original final EIS. But it should thoroughly cover the supplemental proposed project. This SSEIS does not. I have also read the original EIS and the first SEIS mentioned above.

The most obvious shortcoming is the absence of information about the risk of moving the NATO agent and ordinance from Germany to Johnston Island. Earlier in this chapter, I compared the Johnston Island EIS with the LNG proposal in Chapter 4. Neither provides information about extracting the materials from their location and moving them across the ocean. However, there is an important difference between the two proposed actions. Presumably, residents of the Baltimore area and the surrounding mid-Atlantic region would benefit from the natural gas, and there would be some employment, although not much, over the lifespan of the LNG facility. The geography of the benefits would not equal the geography of the risk, but there are some benefits.

The Johnston Island project offers benefits to contractors, and the SSEIS asserts that the Army's presence protects sea turtles from hunters. There are local benefits in the Johnston Island case, to elected officials of Germany who have successfully persuaded the United States to move the agent and munitions from Germany, and to residents who lived in the community surrounding the weapons storage area in Germany. Of course, had there been an accident during the transport of these weapons, the benefit would have been replaced by serious problems for the German government. But there were none. Also, there are benefits to residents of the eight states of the continental United States whose risk was reduced by piloting the incineration technology at Johnston Island. None of these benefits is explicitly discussed in this report, and for me this is a remarkable omission. Perhaps the Army assumed them to be too obvious.

If there were to be any benefits to the Johnston Island area, they are not defined in the SSEIS. Even the soldiers and contractors are not local. Furthermore, the SSEIS does not claim any local benefits.

The most troubling omission is that the movement of these materials from Germany to a remote island in the Pacific stands in clear contrast to policy for the continental United States, which does not permit unitary chemicals weapons and agent to be transported across state boundaries. If the agent and weapons cannot be moved within the United Sates, what is the rationale for moving them across the ocean? The obvious answer is that the proposed action was a political decision.

This interpretation is reinforced by the explicit statement that the set of technologies had to be proven before they could be used in the continental United States. For residents of the Hawaiian Islands and Micronesia governments, it was not a leap of faith to assume that this region, where nuclear weapons were tested, and Agent Orange was burned offshore, was once again a sacrifice zone for the United States. The testimony of regional residents emphasized this point – the people felt like guineas pigs. The environmental justice literature did not emerge until the 1980s, and President Clinton's executive order requiring explicit consideration of environmental justice was not until 1994. I suspect that that this would have been a different document had the decision been made in 1995 rather than 1990, but that is speculation. There is no way around the observation that this action appears to be insensitive to environmental racism and to the history of this area. To the people of this Pacific region, this action was unfair.

The Army did not even consider the impact of a potential accident on local fishing and recreation. Instead, these concerns were dismissed because the additional ordinance would only take 3.5 months to destroy, and the report did not consider the possibility of an accident during transport.

The science and engineering is also limited, compared with what would now be expected. The plume models were state-of-the-art then, but the assumptions used in the modeling lacked sufficient local data. However, it is hard to see how the results could lead to an imminent threat to Hawaii and other populations in the Pacific area. The remoteness of Johnston Island was protective. Better data and models would not have led to evidence to change the decision.

The tone and writing are satisfactory. Indeed, the message to regional residents is remarkably clear. If the writers wanted to hide the message, they certainly did not do a very good job of burying it in facts. The report is relatively short, and easy enough to read. The biggest problem I have is the missed opportunities to explain this decision in the way that Professor Kosson did.

## Comprehensiveness

This document, indeed all three documents, were not comprehensive. The components of the life cycle that are presented are adequately discussed, although, as noted above, that would not be my conclusion if the report had been issued in 1995. By starting the EIS off the coast of Johnston Island and asserting that it would only take 3.5 more months to destroy the nerve gas, the SSEIS was figuratively pre-ordained to find no significant impacts. After all, if the full program takes 3–4 years for the Okinawan ordinance, then 3.5 more months for the NATO agent is a minor addition to the risk. Again, the writers of this document had opportunities to offer more thoughtful explanations than they did.

## Coordination

Judged by their terse responses to the SSEIS, the cooperating federal agencies were supporting the decisions; however, the author (perhaps projecting) reads some frustration in their official responses. The EPA urged the Army to make information available: "Since the publication of the Draft EIS, much additional information concerning the munitions and the proposed program has been declassified and made publicly available. We strongly recommend that the formerly classified information be included in the Final EIS" (US Army 1990b, p. 5).

EPA offered other suggestions about sampling, notably about how the information will be distributed to the public:

> The Final EIS should provide additional information concerning the trial burns under the Toxic Substances Control Act (TSCA) and operational verification testing. This should include the materials incinerated and compounds monitored. . . . In addition, the Final EIS should include data for the stack gas toxicity monitoring program. . . . For work to be completed in the future, the Final EIS should indicate at what point and how the results of these three procedures will become available to the public.
>
> (*ibid.*, p. 8)

EPA also added the following statement about impacts on non-island personnel: "The Final EIS should discuss the potential impacts on other populations (fishing fleets, etc.) at risk in the event of an accidental release" (*ibid.*, p. 10). I found it shocking that these suggestions had to be made by the EPA in a written report; all this seems to be obvious without the need for EPA prompting.

The US DOE has had more than its share of EISs that have drawn heated responses. Frank Bingham, Director of the DOE Environmental Protection Division, offered five comments, many focusing on the report's failure to explain the science and engineering, even including a discussion of public participation and not providing a list of persons, groups, and agencies contacted. Again this author found these terse statements to be surprising insofar as the lead agency would not normally have such obvious omissions. These statements are illustrative of the lack of comprehensive treatment of issues in this SSEIS.

Representatives of the State of Hawaii opposed the proposed action and, along with others, were angry about not having the documents made available to them in a timely manner. The choice of language by Representative Annelle

C. Amaral is quite telling, as it points to a failure to deal with key issues in the document:

> While some may argue that storage of these chemical weapons may pose hazards due to the possibility of their containers developing leaks or through accidents, I would note that the possibility of accidents occurring is much greater every time these containers are handled. Moreover, the possibility of an accident will be infinitely greater if this project is carried through – these containers will have to be shipped half-way around the world, during which trip they will probably have to be handled at least five times: loaded onto a train or truck in Europe, loaded onto a ship, unloaded from the ship to other vehicles at Johnston Atoll, placed in a storage area, picked up from the storage area, and placed in the incinerator facility. That is the minimum of handling I can envision. Add to that the threat caused by normal wear and tear during transport and the possibility of accident, or even sabotage, during transport. Add to this equation the law of human activity, that if anything can go wrong it will go wrong.

> The Army, in its 1983 EIS, stated that there are presently no plans to dispose of chemical agents or munitions on Johnston Island that are not stored there at this time. Now, however, less than a decade later, the Army is attempting to overturn this statement of its intent and purpose in the development of the Johnston Atoll Chemical Agent Disposal System to allow itself to bring materials to Johnston Atoll from the four corners of the world for disposal, if necessary.

> (*ibid.*, pp. 29, 230)

These statements should not have been needed to be raised by a representative of a state government. These issues should have been addressed in the SSEIS.

## Accessibility to other stakeholders

The author concludes that this decision had been made, and no other options were to be considered, because of political and legal decisions. Hence it is hard to understand why the documents were not made available far in advance of the meeting. A conspiratorial theorist would argue that the Army deliberately waited to make this document available to avoid a politically embarrassing confrontation in Hawaii. However, the Army representatives seemed genuinely embarrassed by the charges of not making the documents available. Furthermore, the author has been at meetings in the continental United States where an estimated 500 people were present and used language that was a lot stronger than that used in this Hawaii meeting. The Army chose incineration in those cases, and was under no obligation to change its decision in the

Johnston Island case. This case study clearly demonstrates that access to the documents and process does not imply a different decision. With documents now automatically available on the web, the complaints aired at Johnston Island should not be repeated.

## Fate without an EIS

The decision was a political one, and the SSEIS probably slowed down the implementation, but not by much because of the time-sensitive political decision.

# 6 Savannah River Nuclear Weapons Facility: managing the legacy of the military's nuclear factory

## Introduction

Before the Second World War, the idea of building nuclear weapons was a scientific possibility shared among a relatively small group of physicists (McMillan 2005; Walker 2005). The first nuclear weapons were built and used by the United States to end the resistance of Japan. Subsequently, tens of thousands of nuclear weapons have been built, almost all by the United States (see below) and the former Soviet Union. With proliferation, the United Kingdom, France, China, Pakistan, India, Israel, and North Korea have, or are assumed to have, produced nuclear weapons.

The US government began nuclear weapons production during the 1940s under the "Manhattan" project, and while it maintains thousand of nuclear weapons, the Department of Energy (DOE) has been dismantling unused weapons and managing their by-products (Office of Environmental Management

1995). This chapter is about the use of the EIS process in the DOE's efforts to manage this nuclear defense waste legacy.

I had four reasons for writing a chapter on the nuclear weapons legacy. First, the complexity of managing this legacy is remarkable; I think unprecedented. The United States literally created a distributed nuclear factory at over 130 locations in more than thirty states to design, build, and test nuclear weapons and manage the by-products of these weapons (*ibid.*). The nuclear bomb effort has required unprecedented cooperation among scientists, engineers, and many other professionals at multiple sites. I wanted an EIS that would provide a snapshot of the complex process of managing important pieces of the nuclear waste cycle.

Second, because some of the nuclear and chemical materials are extremely hazardous and will remain so for tens of thousands of years or longer, the science and engineering has been extraordinarily challenging, in many cases requiring first-of-a-kind design, engineering, and evaluation. This type of EIS is the most technically demanding one described in the book. It often contains numerous flow charts, tables and terse technical summaries and jargon, and it represents many individual projects. I tried to focus on the major points and on a few components of the EIS in order to provide some depth. It is important for readers to see for themselves that some EISs are difficult to read, and doubtless difficult to write.

Third, the cost of this national effort at over 100 locations has already been $90–100 billion, and could reach over $300–375 billion (Office of Environmental Management 1995, 1997; interview with Henry Mayer, Executive Director, National Center for Neighborhood and Brownfields Redevelopment, Rutgers University, May 18, 2010). The cost of this waste management program is larger than that of any other government-run waste management program in the United States, and I believe larger than any other in the world, albeit it is not the most serious nuclear-related problem. The construction and operation of projects described in this EIS by far exceed the cost of any other project described in this book. What is notable about these costs is the juxtaposition of high cost and relatively minor near-term risks. Like the destruction of the chemical weapons on Johnston Island in Chapter 5, the risks are relatively small and are driven by legal mandates on the US government, stakeholder agreements with the DOE, and by the US government's ethical commitment to close the circle on these weapons of mass destruction (Carter 1987; Office of Environmental Management 1995). No US resident is being supplied with a road or train for travel, an energy product to heat their home or business, a historical place to visit, water to drink, or another obvious useful service. These EISs represent a set of documents written to address legal mandates and treaties, which implies a different role for project cost and politics than is normally found in business.

Fourth, the larger DOE sites are in relatively remote locations, where there are no immediate neighbors. But these sites are not ignored. Four regions, including Savannah River and Hanford in the United States, have depended on the DOE's weapon production, decommissioning, and waste management programs for their economic health. Accordingly, there is great local concern about DOE's expenditures, and trepidation about them stopping. The public reaction to new waste management projects is not like those I have seen elsewhere. Many strongly support new facilities. For example, a recent study showed that a near majority favored locating new nuclear power plants in the Savannah River region, and more than a majority favored locating new nuclear waste management facilities in this region. Both figures were notably higher than a national sample (Greenberg 2009). In addition, each of the major sites has an officially appointed citizen advisory board, and their reactions to the EISs are interesting (see below for further discussion).

Given these four reasons for wanting a case study of nuclear waste management, I could have chosen from over a dozen national programmatic and site-specific EISs on a variety of subjects. I have worked on nuclear waste management on and off since the 1970s, and so it was important that I not pick any single project that I had studied in detail, and therefore might, consciously or otherwise, have a bias toward or against because of my own research. Instead, I wanted to focus on a major planning document that involved numerous projects.

I chose to focus on one seemingly innocuous final EIS written in the mid-1990s for the Savannah River Site (SRS) in South Carolina. This EIS (US Department of Energy 1995a) illustrates the four reasons I wanted to present an analysis of this legacy management program: complex science and engineering, extremely hazardous materials, high costs, and a notable public involvement process. This site-wide environmental management EIS has embedded within it dozens of individual actions at the site and elsewhere across the DOE nuclear complex. For example, one set of EISs was written for the Defense Waste Processing Facility (DWPF) (US Department of Energy 1995a,b), which takes high-level nuclear waste and mixes it to produce large glass canisters (which look like large metal logs) that immobilize the highly radioactive waste (see below). The DWPF was my first choice for a nuclear waste EIS because it was such an interesting project and EIS. However, it is a very technical document and, more importantly, would not provide the opportunity to examine the full set of issues that this site-wide document presents; furthermore, the DWPF is part of the EIS to be reviewed here.

Before reviewing the specific EIS, I provide context about nuclear weapons and their wastes, and regulation of military nuclear waste. The EIS will be easier to understand in the appropriate context.

## US nuclear weapons and wastes

US nuclear weapons are complicated devices designed to explode only when authorized to do so. This means that they have complex safeguards to prevent accidental or deliberate and malicious detonation. There are a number of basic types of nuclear weapon. So-called "atomic bombs" use high-grade conventional explosives and uranium that has been enriched with fissionable materials. The high-grade explosives trigger fission of the enriched uranium, that is, the enriched uranium atoms split apart causing a chain reaction. US nuclear weapons use the artificial element plutonium as the fissionable material. The explosives momentarily compress a ball of fissionable material in a metal sphere sufficiently to reduce neutron leakage so that the neutrons released during fission generate a "chain reaction" that produces more neutrons in each successive generation (Walker 2005, Greenberg *et al.* 2009).

Modern US nuclear weapons are so-called "thermonuclear" devices. They use conventional high-grade explosives to trigger fission, which then triggers fusion of heavy hydrogen atoms in tritium gas. In other words, atoms are fused together, rather than split apart as they are in fission. Fusion is what powers our Sun. Thermonuclear weapons release energy from both fusion and fission.

To put the nuclear factory and the Savannah River waste management role in a broad context, the Brookings Institution (1998) compiled facts about US nuclear weapons mostly from published US government sources. The United States built more than 70,000 nuclear warheads and bombs, 67,500 missiles, 4680 bombers, and eighteen ballistic missile submarines to carry these weapons. Fissile materials produced were 104 metric tons of plutonium and 994 metric tons of enriched uranium. In 1966, which was the peak year for nuclear weapons in the United States, the nation had 32,193 nuclear weapons and bombs. Stephen Schwartz (1998) estimated that the US government spent about $5.5 trillion on its nuclear weapons and delivery systems. When updated for the changing value of the dollar, this is over $100 billion a year ($100 billion a year from 1943 to 1998 in 1998 dollars; $131 billion a year on average in 2009 dollars). The DOE's environmental management program at the Savannah River has averaged over $1 billion a year for a decade (Greenberg *et al.* 2003; Mayer, interview, 2010, *op. cit.*).

Since the end of the Cold War, the number of weapons has dropped substantially. For example, in July 2007, President George W. Bush set a goal of cutting the number of deployed strategic nuclear weapons in half by 2012. This means that the United States would have between 1700 and 2200 operationally deployed weapons by 2012, the lowest number since 1950, which is the range required under the bilateral agreement between the United States and the former Soviet Union. Both countries have agreed to reduce their weapons-usable plutonium reserves by 34 metric tons (Greenberg *et al.* 2009). On April 8, 2010, President Obama and Russian President Dmitry Medvedev agreed to further reductions in warheads and launchers (Baker and Bilefsky 2010).

The tens of thousands of US nuclear weapons that are no longer part of the assured mutual destruction policy have been taken apart so that they no longer can be used as weapons. Each weapon has thousands of components. The vast majority of these are complex electronics, which are not discussed in this chapter. The actual bomb material, the so-called "physics package," has only several hundred components. The fissionable material, consisting of plutonium that was part of the nuclear weapon, is the focus of environmental management.

The vast majority of the radioactivity in US nuclear waste is in the spent fuel rods from nuclear power generation. However, although the majority of radioactivity is in the spent commercial fuel rods, most of the volume of nuclear waste is military. The United States is no longer producing nuclear weapons, and yet there is a long legacy from decades of nuclear weapons production and associated research. The military waste is not as "hot" as the commercial civilian waste (it was not in reactors very long), but a good deal of it is hazardous. The DOE is processing a substantial amount of defense-related waste at SRS, and the EIS reviewed in this chapter is central to that national effort. In fact, the DWPF at Savannah River is the largest such processing facility of defense waste in the world, at least until the facility at Hanford in the State of Washington is completed.

## Federal laws and regulations

Nearly all of what the EIS in this chapter proposes is required or conditioned by federal laws and regulations. Nuclear waste is categorized by federal laws, regulations, and rules. It is important that the reader understand that the classification does not necessarily correspond directly to hazard levels. Some low-level waste can contain highly radioactive constituents. With that caveat, about 90% of radioactive waste, by volume, is classified as low-level, and only about 0.3% is high-level. However, high-level waste contains about 95% of the total radioactivity of all nuclear waste (Greenberg *et al.* 2009).

The four main categories as they pertain to defense waste at SRS are as follows:

- High-level waste. These wastes are produced by reactions inside nuclear reactors, including those produced by military programs at DOE sites.
- Transuranic waste. This waste contains elements with atomic numbers (number of protons) greater than 92, the atomic number of uranium (hence the term transuranic, or "above uranium".) Most of this waste was produced from defense-related activities. With regard to waste management, transuranic waste includes only waste material that contains transuranic elements with half-lives greater than 20 years and concentrations greater than 100 nanocuries per gram. If the concentrations of the half-lives are below the limits, it is possible for waste to have transuranic elements but not be classified as transuranic waste. At present the United States permanently disposes of transuranic waste generated from military facilities at the Waste Isolation Pilot Plant (WIPP) in New Mexico, near Carlsbad. At the time this EIS was written, WIPP was not open, and there is considerable discussion of transuranic wastes in the EIS.

- Low-level waste. Whatever is not high-level, used fuel from a power plant, transuranic, or by-product material (e.g. uranium mill tailings) is classified as low-level. It includes everything from radioactive garbage (e.g. mops, syringes, and protective gloves) to highly radioactive metals from inside nuclear reactors. Low-level waste is typically stored on-site by licensees, either until it has decayed away and can be disposed of as ordinary trash, or until amounts are large enough for shipment to a low-level waste disposal site in containers approved by the US Department of Transportation. Low-level waste has four subcategories according to activity level and life-span: classes A, B, C, and "greater than class C" (GTCC). On average, class A is the least hazardous, while GTCC is the most hazardous.
- Mixed wastes. Some radioactive wastes have been mixed with non-radioactive hazardous substances, such as organic solvents or other toxic chemicals. Much of this waste (especially the transuranic waste) contains substantial quantities of long-lived radionuclides, such as plutonium-239 and technetium-99. These hazardous components are regulated by the Environmental Protection Agency (EPA).

The EIS described in this chapter was written when the imminent opening of a national geologic repository was assumed. In 1982, Congress passed the Nuclear Waste Policy Act (NWPA, P.L. 97-425), which stipulated disposal of high-level waste in a permanent deep geologic repository (Kubiszewski 2006). First, the defense high-level waste was to be encapsulated in a glass matrix (vitrified) and then placed in canisters (see above). The NWPA required the DOE to look for at least two suitable sites: one in the west, and one east of the Mississippi River. In 1987, the law was amended (National Safety Council 2001; MacFarlane and Ewing 2006) to say that Yucca Mountain in Nevada would be the only nuclear waste repository for high-level waste. The containers would be placed in underground tunnels. The tunnels, also known as drifts, would be about 1000 feet below the surface and, on average, 1000 feet above the water table to minimize natural exposure to moisture.

The DOE submitted a license application to the Nuclear Regulatory Commission in June 2008, requesting authorization to construct a national repository for the disposal of spent nuclear fuel and high-level radioactive waste at Yucca Mountain. The application consisted of a letter describing its purpose, expecting that permanent disposal would begin in 2017. However, this schedule has been delayed by funding limitations, problems obtaining permits, and litigation (for studies and articles see the DOE website on the Mountain repository at www.energy.gov/environment/ocrwm.htm). The Yucca-only decision has become a major political issue. Nevada has fought the Yucca-only policy (Reid 2007). When the SRS impact statement was written, a single repository at Yucca Mountain was assumed. The chances of it ever being used at as a permanent repository are slim, certainly not any time in the near future. While the Yucca site has not received any waste, WIPP, near Carlsbad, New Mexico, began receiving transuranic wastes in 1999 (National Safety Council 2001).

Also, at the time the SRS EIS was prepared, reprocessing of high-level nuclear waste was not assumed for the United States. The Global Nuclear Energy Partnership (GNEP) is a new program to study the impacts of reprocessing spent fuel rods and reuse the uranium, plutonium, and other transuranics for fuel (Greenberg *et al.* 2009). The process would produce nuclear fuel as well as radioactive wastes. Some of the uranium fuel would be exported to other countries as part of a global strategy to prevent proliferation of nuclear weapons. Countries could then have nuclear power plants without establishing their own uranium enrichment or reprocessing facilities, which would reduce the chances of extracting enriched nuclear materials for nuclear weapons. (It is also extremely expensive to obtain fissionable weapons-grade materials in this fashion.) The nuclear fuel could be burned as fuel in existing reactors, or in a new type of reactor to be developed and built in the United States. If it succeeded, the reprocessing and recycling would reduce the total volume and toxicity of highly radioactive wastes (*ibid.*). But GNEP is controversial and has been challenged on economic, environmental, moral, political, public health, technological, and other grounds. GNEP was not considered when the EIS in this chapter was written, and at the time this book was being written, the future of the GNEP concept was undetermined.

Meanwhile, US military high-level waste is being stored primarily at the Hanford, Savannah River, and Idaho sites. The main issue with regard to waste of defense origin is the high-level waste that remains in massive, largely underground storage tanks. These nuclear wastes in tanks were mixed with other substances, substantially complicating every aspect of environmental management. Some of the tanks have leaked, but all of them must be carefully monitored. Waste sometimes is transferred from one tank to another.

## Savannah River Site environment and long-term stewardship

With regard to the local environment, the SRS near Aiken, South Carolina is about 20 miles southwest of Augusta, Georgia (see Figure 6.1). Established in 1950, the 310-square-mile SRS is among the DOE's largest facilities, with 15,000–20,000 employees and a budget well over $1 billion a year (Office of Environmental Management 1995) when this EIS was prepared.

In 1950, E. I. DuPont de Nemours was asked by the federal government to build and operate the facility. Its historic mission in the weapons complex was to produce plutonium (weapons-grade material) and tritium (US Department of Energy 2010). In 1952, the site produced so-called "heavy water," and in 1953, the first production reactor began to operate. In 1956, the facilities that locals often call the "bomb plant" were completed. The defense mission has not disappeared. For example, in 1986, construction of a new tritium facility began.

While the bomb mission was prominent, other activities were occurring. In 1963, spent nuclear fuels from other sites were received. In 1972, the site was

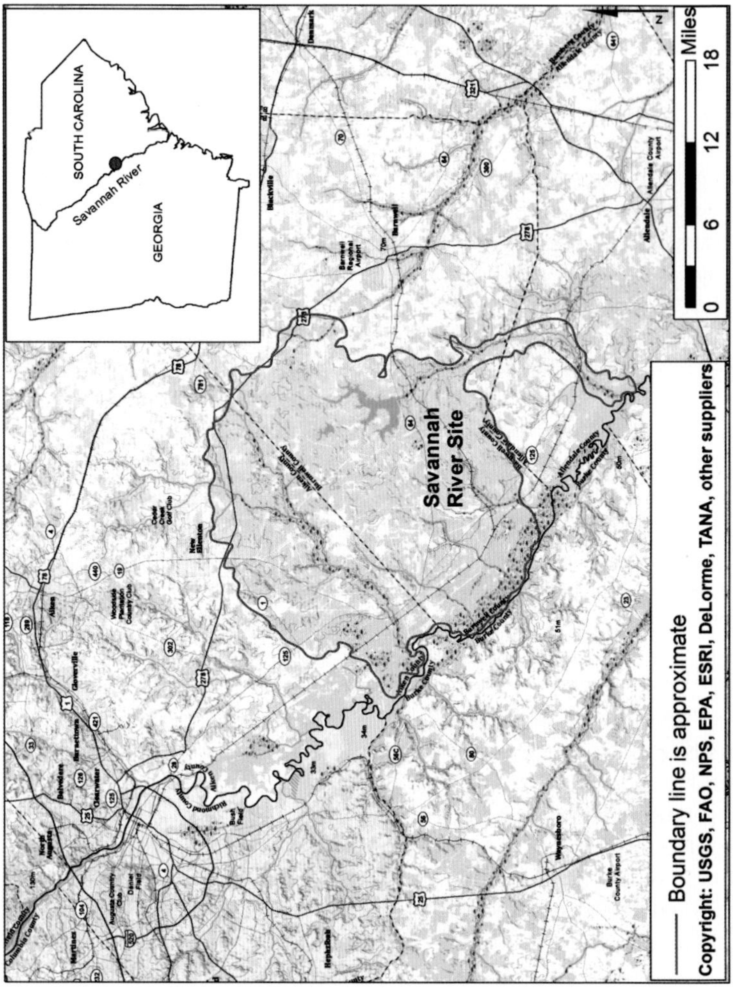

**Figure 6.1**

Savannah River Site

designated as the first National Environmental Research Park. In 1981, the environmental cleanup program began. This included shutting down of reactors, constructing treatment and burial facilities, and starting construction of the DWPF.

When the Cold War ended, SRS's role in the DOE Complex changed primarily to waste management and ecological research. SRS's role has continued to change depending on national needs; it may take on new military roles, and is currently being considered as a location for an energy park that would pursue science in solar energy, wind, biomass, coal, and nuclear energy (Gilbertson 2009).

The DOE's major sites in Idaho, Nevada, South Carolina, New Mexico, Tennessee, and Washington are surrounded by prairies, wetlands, deserts, and forests. Roughly 10% of the land is contaminated with radioactive materials released from the facilities, and there is also some chemical contamination from site operations (Burger 2000; Burger *et al.* 2003). Roughly 90% of the land is not contaminated, and usually represents the few remnants of old-growth forests and other vegetation in the region that have not been disturbed for more than half a century. SRS is illustrative, containing large tracts of forested areas. Almost 80% of the land was deliberately left as buffer areas. In 1997, the author wrote an article that characterized the site as "bombs and butterflies" (Greenberg *et al.* 1997), although another commentator felt that "mosquitoes" was a more accurate descriptor than "butterflies".

The active or formerly active parts of these sites are so far from population concentrations that the hazards they contain pose relatively little risk to people living in the region, although worker risk cannot be discounted. Radiation and chemical exposures to the animals and plants can be a problem. An exposure could occur, for example, when an animal ingests, absorbs, or breathes a radionuclide, or when a plant draws a contaminant out of the soil. The effect varies in accordance with the amount of the dose, length of exposure, and vulnerability of individual organisms to damage. It is important to know that the DOE's mission includes environmental restoration, and DOE's stewardship program includes ecosystem management and integration of economic, ecological, social, and cultural factors in land-use decisions (DOE Order 430.1). When the DOE prepares an EIS for SRS, or for another of the major sites, it includes that ecological responsibility.

Major concerns related to the impact of nuclear materials are exposures to top-level predators (e.g. eagles and hawks) through ingestion, and the health and safety of wildlife populations that may be eaten, hunted, photographed, or viewed, or that are just part of the environment. Threatened or endangered species are always of special concern. Stripping the top layer of soil to reduce radiological contamination can harm existing animal and plant populations more than the radiation levels that existed before remediation (Burger 2000; Burger *et al.* 2003).

Remediation described in an EIS can destroy intact ecosystems that will never recover because of the degree of soil or other disruption. Remediation could be so extensive, for example, that it would destroy the seed bank and allow invasive species to move in, eliminating native species. Also, intrusion will mean movement of nuisance and other animals out of the area to escape remediation, perhaps to nearby suburban areas.

While we may not see what goes on in water bodies, aquatic animals are particularly at risk because contaminants move quickly through water. Also, some animals that do not move far from the contaminant source, or spend more time in the actual contaminated medium (fish in water), will have higher exposure. For many species, leaving waste sites alone would often be the best remedy. Cleaning up contaminated sites can re-expose animals by disrupting groundwater, water, sediment, and soil, potentially releasing contaminants back into the environment.

Ecological risks from radioactive contaminants will decrease if there is no new contaminant release because radioactive substances gradually decay. Yet the risk of some radionuclides will not decrease significantly for many decades or centuries. If action must be taken, the best approach is to avoid new roads being built through the system, and to disrupt the soil, trees, water, and the rest of the ecosystem as little as possible. DOE site-wide EISs must grapple with the dilemma of balancing the removal of hazards and disrupting ecosystems. The document does not provide much context for the non-expert about these trade-offs and their implications for on-site ecosystems.

Some radioactive contaminants with short half-lives will rapidly naturally decay into benign and other not-so-benign substances. However, some of the long-lived contaminants will be radioactive for tens of thousands of years or more. In addition, polychlorinated biphenyls (PCBs), solvents, and heavy metals do not decay much and can remain hazardous in perpetuity. To ensure that human health is protected for many generations, long-term stewardship must be carried out at sites like SRS even after active remediation (Greenberg *et al.* 2009). A remediated site could be cleaned up and even closed, but contain hazards in the form of stored or buried wastes, entombed facilities, such as reactors, or residual contaminants that are left in place.

Engineered controls for this purpose include physical containments around landfills, vaults, tank farms, or other waste units. They include operating, maintaining, inspecting, and monitoring caps, erosion control systems, environmental sampling, wells, and pump-and-treat groundwater remediation systems.

Institutional controls are legal tools to reduce exposure by ensuring that land- and water-use restrictions are maintained. They include restrictions on land or water use, well-drilling prohibitions, deed notices, easements or other legal advisories or measures, and long-term information management. Access obstacles such as fencing, markers, and signs can be considered passive controls. Federal ownership in perpetuity is itself an institutional control. The reader would not necessarily know that these controls are in place, assumed, or required by reading this document.

# The final Savannah River waste management EIS, 1995

This EIS concerns the management of legacy and newly created high-level radioactive waste, low-level radioactive waste, hazardous and nonradioactive waste, radioactive and hazardous waste (mixed), and transuranic wastes at SRS. It does not consider domestic (sanitary waste) or domestic or foreign spent nuclear fuel. The wastes it focuses on must be managed to protect human health and the environment, and be in compliance with regulations, yet the choices are expected to be as cost-effective as possible.

The complexity of this EIS cannot be understated, and it is a reflection of the site itself. When the US government acquired the land in 1951, 60% of the 310 square miles of land area was forest and 40% was cropland and pasture (US Department of Energy 1995a). Almost 60 years later, about 69% was upland forest (predominantly pine) managed by the US Forest Service, 22% was wetlands and water, and 9% was waste management and other activities.

Almost all of the land that is used by the DOE for production, waste management, and administration is concentrated in a core area (Figures 6.2 and 6.3). Outside this developed area, as noted earlier, the site is an ecological treasure, a rare place where humans have had little impact for at least 60 years. Satellite images show a massive forested area with a few open spots. Several roads run through the site. Important for the EIS, the site's waste management operations are influenced by national-level decisions about weapons production and decommissioning, as well as waste management. Readers looking down at the site from above would see three kinds of activity that involve waste management focused in the inner area. One set of activities is ongoing operations related to chemical separation and processing of waste located in the high-level waste tanks. A second set of activities is decontamination and decommissioning of hundreds of buildings. The third major environmental management-related activity is environmental restoration. Each of these is central to decision-making about the site and is reflected in this EIS. In addition, activities at other sites across the DOE Complex impact SRS and are part of this EIS.

This EIS lists sixteen NEPA-related reviews issued between 1987 and 1995 that influence what is presented in this EIS. Five of the sixteen were DOE-wide assessments about tritium supply for nuclear weapons and recycling; nuclear weapons nonproliferation about spent nuclear fuel from foreign reactors; nuclear stockpile stewardship; and two others. The remainder were SRS-specific EISs and initial environmental assessments, including groundwater protection, incineration, mixed waste treatment, the DWPF, and plutonium management.

The impact of nonlocal decisions on waste management at SRS is illustrated by the Pantex and Idaho EISs summarized in this volume. Pantex, located near Amarillo, Texas, assembles and decommissions nuclear weapons. The EIS for the continued operation of Pantex influences the type and amount of waste

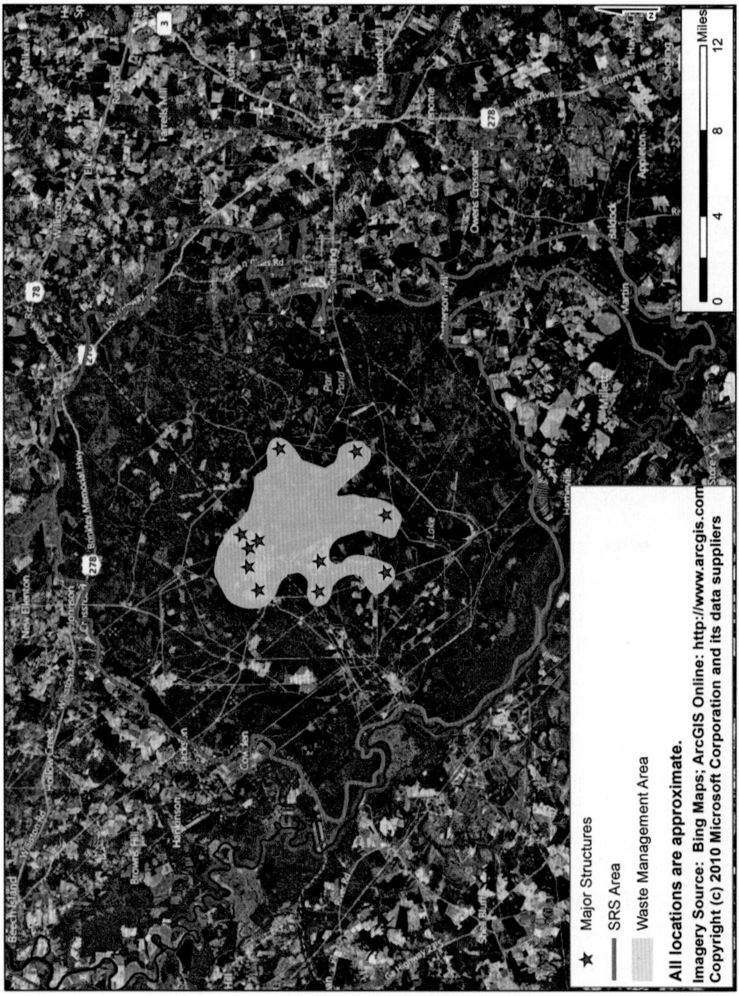

**Figure 6.2**

Savannah River Waste Management Area

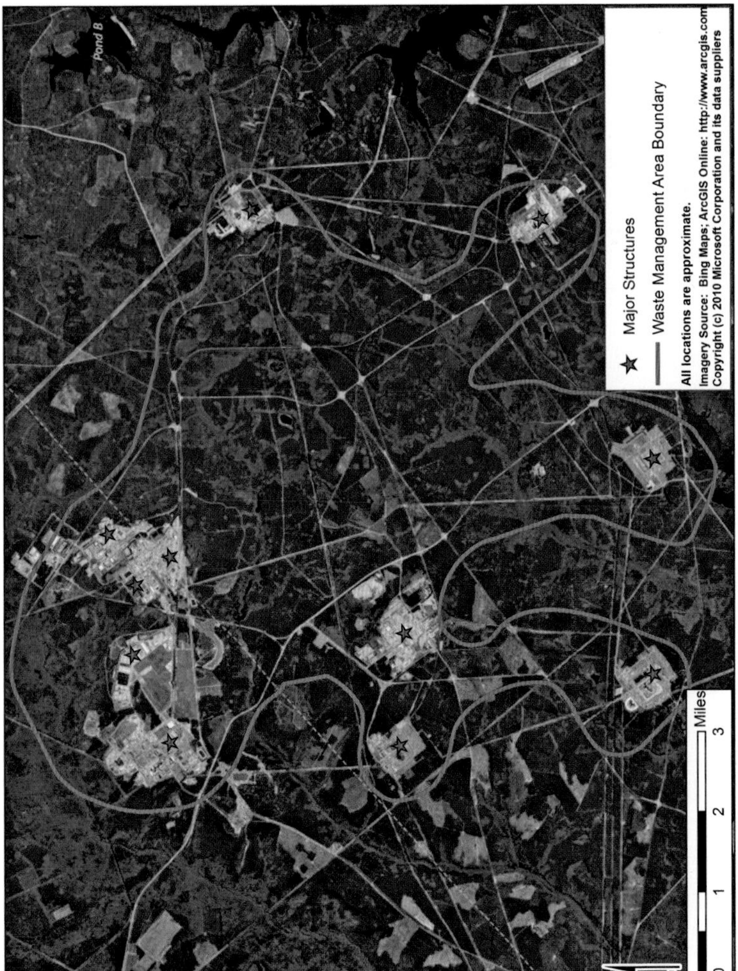

Legend:
★ Major Structures
— Waste Management Area Boundary

All locations are approximate.
Imagery Source: Bing Maps: ArcGIS Online: http://www.arcgis.com
Copyright (c) 2010 Microsoft Corporation and its data suppliers

Pond B

0   1   2   3
Miles

N

**Figure 6.3**

Major developed areas

that would come to SRS. The Idaho nuclear facility programs for spent nuclear fuel are also part of the consideration of this SRS waste management EIS, as was the WIPP facility for transuranic wastes.

My key point is that, in response to federal mandates, the DOE created proposed options to manage multiple sources of waste that were created in the past, are still to be created, and involve transport of products across the United States to and from the site. The author views this site as a giant chess board with pieces moved around at various locations by chess masters located at the site, Washington DC, and other DOE facilities.

## Four options for the Savannah River 1995 EIS

This EIS offers a no-action alternative and three alternatives for managing nuclear waste for the next 30 years (1995–2024). The options are different from any of the other alternatives presented in this volume. In the other chapters, the alternative action was relatively distinct, although some alternatives had overlapping activities. In this case, all four options have overlapping actions, and this EIS has no real no-action alternative. The four alternatives are labeled "no-action," "limited treatment," "moderate treatment," and "extensive treatment."

All of these alternatives have implications that can be measured by radionuclide releases and storage, exposure to workers, the public, and ecosystems, jobs, and in other ways. In order to begin with a uniform and relatively easily comprehensible summary metric, I have provided a summary table of new structures that the document says will be required under the various alternatives (Table 6.1).

### "No-action" option

The "no-action" option does not literally mean no action will be taken. It does mean a continuation of current practices, because a true no-action alternative would violate the DOE's requirements to manage the wastes. This no-action alternative continues past practices of constructing facilities to store new radioactive wastes. The no-action alternative would leave both transuranic and mixed waste untreated, in essence, in storage. This would mean that the DOE would not meet its legal agreements.

Beginning with structures that would exist with all four options, the DOE would construct vaults for the disposal of low-level waste and for hazards and mixed wastes. It would also include a treatment facility, a long-lived water storage structure, a high-level waste evaporator, and a new waste-transfer facility. The no action alternative, like the other alternatives, also relies on the DWPF to convert high-level wastes into vitrified containers.

**Table 6.1** Summary of new waste management facilities proposed by four alternatives and three waste forecasts*

| Facility | Preferred, moderate | | | Minimum | | | Maximum | | | No-action |
|---|---|---|---|---|---|---|---|---|---|---|
| | min | exp | max | min | exp | max | min | exp | max | |
| Storage buildings for long-lived low-level waste | 7 | 24 | 34 | 7 | 24 | 34 | 7 | 24 | 34 | 24 |
| Mixed waste storage buildings | 39 | 79 | 652 | 45 | 79 | 757 | 39 | 79 | 652 | 291 |
| Transuranic and alpha waste storage pads | 2 | 10 | 1168 | 3 | 12 | 1168 | 2 | 11 | 1166 | 19 |
| Organic waste tanks in S, E, and aqueous waste tanks in E-area | 0 | | | 0 | | | 0 | | | 73 |
| Shallow land disposal trenches | 37 | 58 | 371 | 25 | 73 | 644 | 45 | 123 | 576 | 29 |
| Low-activity waste vaults | 1 | 1 | 8 | 9 | 12 | 31 | 2 | 2 | 8 | 10 |
| Intermediate-level waste vaults | 2 | 5 | 9 | 2 | 5 | 31 | 1 | 2 | 3 | 5 |
| RCRA-permitted disposal facilities | 20 | 21 | 96 | 21 | 61 | 347 | 10 | 40 | 111 | 1 |

* Waste forecast: min = minimum; exp = expected; max = maximum.

Adapted from US Department of Energy 1995a, EIS-217, Table S-1, p. S-21.

Table 6.1 shows that the DOE would be required to build seventy-three new waste tanks in three areas of the site that would not be required with the other options (see the no action column of Table 6.1). It would also require building shallow land disposal trenches, low-activity and intermediate level waste vaults, and a Resource Conservation and Recovery Act (RCRA)-permitted disposal facility.

Before describing the other alternatives, two other issues must be noted. All of these options, according to the document, would protect public health and the environment. In other words, the DOE's legal mandates are satisfied by any of the three.

The DOE focused on five criteria when it evaluated the options:

- waste processing variables, such as achievable volume reduction, secondary waste generated, and efficacy of decontamination and decommissioning;
- engineering variables, such as effectiveness of technology, its maturity, and needed maintenance;
- safety, public health, and environmental impacts, including worker, public, environmental, and transportation risk;
- public acceptance and political consequences, including regulatory requirements, permits, and schedule;
- cost-effectiveness in the near-term and over the next 30 years.

Local site managers, as noted above, do not have complete control over what they will do on the site. Analysts recognized this complication, and thus estimated the minimum, expected, and maximum amount of waste the site could receive for the following five waste categories: liquid high-level radioactive waste; low-level radioactive waste; hazardous non-radioactive waste; mixed (radioactive and non-radioactive hazardous) waste; and transuranic waste.

Hence, rather than looking at four options, the reader looks at ten (one no-action; and three each for the limited, preferred, and extensive options). Because of the complicated options, I have chosen to present the preferred option last in the text rather than first.

## Limited treatment option

This option meets required mandates, but not much more. For example, low-level waste would be treated by existing compactors already in place before storage. Hazardous waste would be recycled, then sent off-site for treatment and disposal, or incinerated in the Consolidated Incineration Facility (CIF).

The waste forms would be safe, but not represent state-of-the-art treatment. A negative implication is that there would be more waste generated. Yet an advantage is that potential worker and public exposure would decrease in the short run because there would be less handling of the waste. Table 6.1 shows that this approach requires more waste storage vault and RCRA-permitted disposal facilities than the other two options.

## Extensive treatment option

In contrast to the limited treatment option, this one would require far more extensive treatment of wastes to reduce their toxicity and volume and create stable waste forms that would be difficult to move. The long-term toxicity impacts would be reduced, but the trade-off is that there inevitably would be more exposure to workers and possibly the surrounding population due to additional handling and processing of the waste. This option means more reliance on the incineration facility to burn wastes, repacking and then storage or shipment of waste. Table 6.1 shows more reliance on shallow land disposal trenches, but what is not shown in the table is the greater reliance on vitrification for low-level waste at SRS and even other hazardous wastes.

## Moderate treatment – the preferred alternative

Most moderate treatment options can be seen as a way of reaching a politically and/or economically acceptable solution. Arguably, the same is true in this case. But there are health and safety reasons for this choice. The EIS takes a position that the DOE chooses to concentrate its resources on the wastes it considers most likely to impact human and environmental health, and will place less emphasis on treatment for less hazardous wastes. A good example is transuranic wastes. Under the extensive treatment scenario, all transuranic wastes are to be vitrified. Under the least treatment option, they would all be repackaged in accordance with federal law. The preferred option distinguishes between plutonium-238 and highly radioactive plutonium-239 components of transuranic wastes on the one hand, and all the other transuranic forms on the other hand. The first two would be vitrified and the remainder repackaged. In essence, this is a risk-based approach to allocating limited waste management resources.

Public reports of the EIS focus on the DOE's choice of the moderate treatment option. The most obvious differences in Table 6.1 are between the minimum and maximum waste inventories that would have to be managed. The number of long-lived low- and high-level waste storage buildings does not vary among the alternatives. But the number of mixed waste and transuranic storage pads is markedly greater if the site generates and receives additional waste products, as do the number of shallow land disposal tranches, low-activity and intermediate waste vaults, and RCRA-permitted disposal facilities. This distinction is clearly highlighted in Table 6.1.

Behind these notable differences in the impacts of the options are policy-related assumptions about what is going to happen both on- and off-site. The

DOE's analysts took 1993 and 1994 data and made assumptions about what could conceivably happen. Space does not permit a full recitation of what the waste forecasts assume. For example, the maximum waste forecasts make different assumptions about aluminum-clad spent nuclear fuel coming to SRS for processing from Idaho; about plutonium and tritium from Pantex coming to SRS; and nine other questions. Depending upon the answers to these questions, SRS could have a great deal more waste, which was envisioned in the maximum waste forecast. For example, the expected waste forecast for the years 2000–24 assumes that 182 facilities inside the central area will be gutted and 423 outside the area will be taken down to their foundation. In strong contrast, the maximum waste results assume that all 182 in the central area will be taken to foundation and all 423 outside the area will be taken to "greenfield." The difference between gutting, taking to foundation, and removing implies a lot more waste being moved around the facility.

A more detailed example is how the three options plan to deal with the residuals of spills, so-called "spill units." The minimum forecast assumes that forty of the 134 spill units (30%) would have their wastes removed. In contrast, the maximum waste forecast assumes all 134 would have their wastes removed. The expected waste forecast assumes that half (sixty-seven) would have their waste removed.

Before highlighting some of the impacts, it should be noted that the 30-year study period for this plan leaves the site with ongoing environmental management operations that will last well into the twenty-first century. Table 6.2 (opposite) lists the facilities that manage the high-level nuclear waste.

## Environmental impacts considered: some illustrations

The 310-square-mile SRS site sits on the Atlantic Coastal Plain, and includes parts of Aiken, Allendale, and Barnwell Counties in South Carolina. The impact sections examine potential environmental effects during both the construction and operations of proposed new facilities. The report examines impacts on air, water, animals, and plants, and on worker and public health for those who live in the area. The report also includes presentations about economic and social impacts.

None of the environmental conditions is striking. Impacts vary much more by amount of waste received than by waste management options (no-action and three alternatives). The key to understanding the message is that the waste forecasts mostly depend on the extent of decontamination, decommissioning, and restoration of the site. Rather than go through every category, I focus here on some of the key impacts that were addressed (see Table 6.3, p. 158).

**Table 6.2** Facilities that would operate beyond planning period, 1995–2024

| Type | Function |
|---|---|
| Defense waste-processing facility | Vitrifies high-level radioactive waste |
| Z-area saltstone manufacturing and disposal facility | Saltcrete processing and disposal |
| F/H-area effluent treatment facility | Treatment of routine process effluent and wastewater |
| In-tank precipitation | Removal of radionuclides from highly radioactive salt solution |
| Savannah River Technology Center | Research and development |
| Replacement tritium facility | Separates tritium from targets |
| Type III liquid high-level waste tanks | Storage of liquid high-level waste, sludge and saltcake |
| Consolidated incineration facility (stops operating under maximum treatment alternative in 2006) | Destroys selective radioactive and other hazardous wastes |
| New special recovery facility of 221 FB-line | Recovery of plutonium scrap |
| Powerhouse, water treatment and support facilities, analytical laboratories | Produce on-site energy, treatment of powerhouse effluent and laboratory services |

Adapted from US Department of Energy 1995a, EIS-217, Table 2-3, p. 2-21.

## Land-use impacts

Table 6.1 shows the difference in number of facilities, which is reflected in land-use impacts. The no-action alternative would require an estimated 160 more acres (quarter of a square mile). This compares with 107, 157, and 1010 acres (the last is 1.6 square miles) for the preferred minimum, expected, and maximum waste forecast estimates, respectively. The document indicated that each forecast is consistent with the site land-use plan, which is to concentrate development in the central area (see Figure 6.2).

## Ecological impacts

Under the no-action alternative, 160 acres of forest would be required. The requirement for the preferred option would be 117 more acres under the

**Table 6.3** Environmental impacts considered for Savannah River Nuclear Weapons Facility

| | |
|---|---|
| Geologic resources | |
| Groundwater resources | |
| Surface water resources | |
| Air resources | Construction<br>Operations, including nonradiological<br>Air emissions impacts<br>Radiological air emission impacts |
| Ecological resources | |
| Land uses | |
| Socioeconomics | Construction<br>Operations |
| Cultural resources | |
| Aesthetics and scenic resources | |
| Traffic and transportation | Traffic<br>Transportation |
| Occupational and public health | Occupational health and safety<br>Public health and safety<br>Environmental justice |
| Facility accidents | |
| Unavoidable adverse impacts and irreversible or irretrievable commitments of resources | |
| Cumulative impacts | Existing facilities,<br>New and proposed facilities or programs<br>List of full set of impacts on land, water, air, socioeconomic, transportation, occupation and public health |

Adapted from US Department of Energy 1995a, p. 4-191-B.

expected waste forecast. Under the minimum waste forecast, only 90 acres would be required. The notable difference is that 960 acres (1.5 square miles) would be required for the preferred option and maximum waste combination. This represents less than 1% of the forested area on the site. Yet threatened and endangered species and wetlands could be affected, although the DOE presumably will do surveys to avoid this. Avoidance of specific ecologically important areas and other management options would also reduce impact on endangered species that inhabit these historical forests.

Large and mobile animal species inhabiting the undeveloped portions of the site (e.g. fox, raccoon, deer) should be capable of avoiding the construction equipment. However, small and relatively immobile species (small mammals, reptiles, amphibians) would not. Dispersion of the survivors would, in turn, pressure other species in adjacent areas. The net result is fewer total species and possibility less diversity, despite the very small magnitude of these impacts. Timing of the land clearing would be important. The DOE would try to avoid the spring and summer, when the impacts on nests and breeding would be maximum. Impact on wetlands and streams could be minimized if the DOE required the installation of erosion control and used best practices.

## Socioeconomic impacts

The Savannah River region is heavily dependent on the DOE's activities. In an earlier study, the author and colleagues (Greenberg *et al.* 2003) estimated that 16% of the gross regional economic product in counties near the site came directly from site activities. In the nearest areas, the proportion is more than half the product. In a large urban area such as Atlanta, 170 miles to the west of the site, SRS would account for less than 1% of the local economy. Accordingly, economic impact is important in site-specific EISs when the region is otherwise rural, and SRS has experienced serious regional economic recessions when the DOE shrunk the workforce while the United States as a whole was growing. On the other hand, the area has been less impacted than places such as Detroit and other private manufacturing centers when national recessions have occurred (Greenberg *et al.* 2003). Reflecting its economic importance, the local media have made this and similar rural sites a major priority (Greenberg *et al.* 2008). Any possibility of new jobs and regional income is important to many people.

The economic impacts are based on the estimated construction and operations personnel required to implement the remedial options described in the EIS. Impacts on socioeconomic resources can be evaluated by examining the potential effects from both the construction and operation of each waste management alternative on factors such as employment, income, population, and community resources in the SRS region.

Over 100 jobs are estimated to be required for constructing the no-action alternative facilities in the peak construction year. Operations employment is much higher, about 2450 jobs a year during the period 2003–24. This is about 12% of the mid-1990s SRS employment. While this is a sizeable number of jobs, the site was expected to lose over 4000 jobs (from 20,000 to 16,000). Hence DOE expects that these added jobs would be filled through the reassignment of existing workers.

With regard to job creation, the maximum waste forecast presents a notable difference. The maximum peak employment estimate is 330 jobs during the period 2003–05. Operations employment associated with this maximum forecast was estimated to be 10,010 from 2002–05. This is about half of DOE's forecasted regional employment in 2005. The report assumes that half of workers could come from the existing labor force and the other half would be added. If this had occurred, it would have meant more people, more income, and indirect benefits to the local economy. There would be more demand on schools and other services, but more money to pay for these services, and as a result a net addition to the regional economy of between 1.5% and 3%.

## Groundwater impacts

One of my issues with the EIS is that parts of the document are not readable to anyone without a technical background, multiple documents in front of them, and/or a technical glossary (see below for a lengthier discussion). I illustrate this with direct quotes from the report about groundwater. I chose groundwater because water resources are always a major public concern. With regard to the no-action alternative, I quote from several pages that focus on the use of trenches and vaults to contain radioactive wastes:

> The disposal of stabilized waste forms (ashcrete, glass) in slit trenches was not evaluated in the Radiological Performance Assessment and is subject to completion of performance assessments and demonstration of compliance with performance objectives required by DOE Order 5820.2A ("Radioactive Waste Management"). Therefore, DOE was unable to base an analysis of stabilized waste in slit trenches on the Radiological Performance Assessment. The analysis presented in the draft EIS did not account for the reduced mobility of stabilized waste forms in slit trenches. The final EIS assumes that releases from these wastes in slit trenches would not exceed the performance objectives specified by DOE Order 5820.2A. As a result of the modified assessment approach, exceedances for uranium and plutonium isotopes identified in the draft EIS under some alternatives and waste forecasts are no longer predicted to occur. DOE would re-evaluate the performance assessment and, if necessary, adjust either the waste acceptance criteria or the inventory limit for the storage or disposal units to ensure compliance with these criteria, or standards which may become applicable in the future. The results of applying this assessment methodology to the different storage and disposal facilities are presented below.
>
> (US Department of Energy 1995a, pp. 4-10,4-11)

This 193-word example is typical of this EIS and many others prepared by the DOE; that is, it is full of technical language, reference to other sections of the document, and other reports. The author is not accusing the writers of providing inaccurate information or of deliberate obfuscation with technical complexity. However, relatively few people would have the expertise to understand these sections and, even if they did, the author doubts if they would have the patience to read hundreds of pages of text. The appendices referred to are even more technically oriented.

The essence of the message about the trenches and vaults is that institutional controls are essential to make sure that the waste vaults and trenches are maintained. If institutional controls were abandoned and the equipment and vaults were permitted to degrade, there eventually would be leaks into the surrounding land and then to groundwater. The extent of the impact is uncertain because of limited data and modeling.


## Worker and public health impacts

This EIS considers various potential exposures and health effects among workers and the surrounding public. The DOE calculated chronic emissions and accident scenario exposures and consequences. These analyses are some of the more interesting science-based work in the EIS. This site as a whole and the new facilities are very remote from the nearest residential concentration some 25 miles away. Accordingly, the off-site human health impacts are very small, and in essence are not measurable. I will present the worker latent cancer estimates. I picked cancer as the major concern because there is a great deal of radiation on these sites and cancer is such a prevalent disease, making it difficult to distinguish site-related exposures from other contributing factors. I note that some DOE sites have a legacy of potentially toxic beryllium, asbestos, and other metal contamination. But these vary by site, and are not serious problems at every major DOE site in comparison with radiation, which is a consistent concern.

With regard to the no-action alternative, exposures result from handling effluent destined for treatment facilities, waste tank farms, and storage pads. The DOE used industry standard codes and assumptions about exposures at various distances from the facilities. DOE regulations (10 CFR 835) require that annual doses to individual workers not exceed 5 rem per year (roentgen equivalent in man, a measure of the effects of ionizing radiation on humans). The DOE assumed that exposure to the maximally exposed involved worker at SRS would not exceed 0.8 rem per year due to engineering and administrative controls. Using this set of assumptions, the DOE estimated latent cancer fatalities. The probability that the average involved worker would develop a fatal cancer sometime during his/her lifetime as the result of a single year's exposure to waste management-generated radiation would be approximately $1.0 \times 10^{-5}$, or 1 in 100,000. For the worker exposed to the administrative limit (0.8 rem), the probability of developing a fatal cancer sometime in his lifetime as a result of a single year's exposure would be $3.2 \times 10^{-4}$, or approximately 3 in 10,000. For the total involved site workforce, the collective radiation dose could produce up to 0.022 additional fatal cancers as the result of a single year's exposure; over the 30-year period, the report summarizes that the involved workers could have 0.65 additional fatal cancer as a result of exposure.

The preferred alternative presented in the EIS implies the operation of the CIF, vitrification facilities, a mixed and hazardous waste containment building, the mobile soil sort facility, compaction facilities, and the transuranic waste characterization/certification facility. Emissions from these facilities would slightly increase potential adverse human health impacts compared with the no-action alternative for the three waste forecasts. The report examined potential releases and controls to estimate quantities of radionuclides released by each process.

Adding across these activities, the DOE estimated a dose of 0.037 rem per year, which is below the SRS administrative guideline of 0.8 rem per year. The probabilities and projected numbers of fatal cancers from 30 years of waste management operations for the preferred alternative leads to an individual worker to have a 1 in 44,000 probability of developing a fatal cancer due to exposure to SRS waste management activities. Given a workforce of 2154 workers (see above), the estimate is one additional fatal cancer from the 30 years of waste management activities considered in this EIS.

The estimate for the minimum waste forecast is almost identical. The larger amount of handling associated with the maximum waste forecast leads to a higher probability of a worker contracting a fatal cancer as the result of a 30-year occupational exposure to radiation. But "higher" is not much higher. The EIS estimates that two people in the workforce of 2501 could develop a fatal cancer sometime during their lifetimes as the result of a 30-year exposure.

The report places these estimates in context, noting that in the US, 23.5% of the population died of cancer during this period. This means that, if this percentage of deaths from cancer remained constant, 491 workers in a work-force of 2088 involved workers would normally be expected to die of cancer. This is not a false statement; however, it is not an entirely accurate statement because it fails to take into account the "healthy worker" effect, which shows that a smaller percentage of employed people would be expected to die from cancer than the general population, because the disabled and ill are less likely to be employed. Also, comparisons of this type often offend people who reject the idea of accepting any additional cancer burden that could be avoided. In essence, however, the estimates imply that that worker cancer effects would be undetectable.

Recapitulating, the limited treatment alternative requires more disposal capacity and facilities, coupled with more sophisticated methods of contain-ment (more vaults and less shallow land disposal), because this option does not reduce or immobilize wastes to the extent of the other options. Yet the other options also require many of the same facilities, but fewer, with different emphases. The alternative involving extensive treatment would produce higher operations-related impacts than those in the alternative involving limited treatment, because more handling and processing of wastes generally produces more emissions and greater worker exposure. The moderate treat-ment alternative is a hybrid that uses options from the limited and extensive options, and produces impacts that fall between the two. The no-action alternative would require more storage facilities than the other three alternatives. Mixed and transuranic wastes would not be treated or disposed of during the three-decade period. This policy would increase the likelihood of health and environmental impacts, including accidents and worker radiological exposure, compared with the other alternatives. In essence, assuming no new technologies, the impacts would be deferred under the no-action alternative. Overall, impacts are small for each of the alternatives because the site is remote, and wastes are already heavily protected by multi-billion dollar engineering structures and systems and a well-paid and dedicated labor force.

## Public reactions

Public participation at this site is different than one would find in response to the overwhelming majority of EISs. Nine of the DOE's largest sites have site-specific advisory boards (SSABs), including Savannah River (Office of Environmental Management 2010a,b). The boards provide the Assistant Secretary for Environmental Management and the site managers with advice, information and recommendations about waste management, technology use, and site restoration options. The DOE views the SSABs as official public representatives and as a way of building public trust.

Members are chosen with regional demographics in mind. As of June 2010, the SRS SSAB had twenty-five members, including eleven women and seven African Americans. The SRS full board has monthly meetings, and its committees meet as needed. All of their minutes are published on the DOE's website, which allowed the author to read those for 1995, which is when this EIS was finalized.

The waste management EIS was mentioned in the January 1995 meeting by Tom Heenan, former assistant manager for environmental quality at SRS, and again by him in the September meeting, when he stated that SRS had chosen a middle path. These were the only notations about this EIS. However, this does not mean that the SSAB was not interested. In fact, throughout the year, the minutes present discussions of almost every major element in part of the waste management EIS. These discussions include high-level waste, the DWPF, groundwater contamination, on-site land-use plans, transuranic waste, receipt of foreign spent fuel, clean-up of contaminated basins on the site, options regarding high-level waste tanks, transportation of nuclear materials through South Carolina, and permissible recreation on the site. At almost every meeting, the Board discussed the pressure on the site's environmental management budget and expressed their concern that the site was not receiving its appropriate share of the DOE environmental management budget.

At some meetings, DOE officials and citizens from the surrounding area praised the SRS SSAB for representing citizen interests. This is not to say that every meeting was free of friction and controversy. After several heated exchanges among members, for example, the Board discussed what they called etiquette. They also discussed if Board members were representative of the surrounding communities, and how many meetings could be missed before someone was asked to resign from the Board. The people who serve on this Board, and others that the author has visited, take their responsibility to the community seriously.

With regard to the process for this EIS, the DOE completed the draft EIS and the EPA published a notice of availability for the document in the *Federal Register*, which started the period of public comment on the draft EIS. Following standard practice, the DOE accepted and responded to letters, telephone messages, faxes, and presentations at public hearings near the site in Barnwell, South Carolina; Columbia, South Carolina; North Augusta, South Carolina; Savannah, Georgia; Beaufort, South Carolina; and Hilton Head, South Carolina. This is quite a few hearings within 100 miles of the site. Yet the DOE received only ten letters and heard statements from five people.

Remarks focused on the CIF; impacts of the wastes on pubic health; and public participation. Government comments primarily sought clarification and/or indicated no opposition to the EIS. The EPA endorsed the proposed action in its response and asked for more information. As previously noted (Chapter 1), this is a common response. William Lawless of the Savannah River SSAB posed several questions at the public meetings.

By reading the letters and documents, I found that that the public's concerns were related to this EIS, but were more focused on issues raised in other EISs about this site and the DOE complex. The DOE responded to all the points and indicated that some points were outside the scope of this EIS, and that they would forward the questions to the appropriate groups. For example, one concern was decomposition of organic materials present in low-level wastes. The testifier suggested that the incinerator ash be vitrified, and that buried contaminated metals be retrieved and processed by smelting before sale or reburial. The DOE respondent replied that these ideas would be consistent with the extensive treatment alternative, but that the EIS does not establish the level of restoration across the site. This, he noted, awaited an agreement between the DOE, EPA, and State of South Carolina.

Concern was expressed about the underlying science and validation of accident scenarios in the EIS. The DOE responded that they used the experiences of similar facilities elsewhere to estimate impacts, and they noted that independent peer review was employed to review their analysis of, for example, flood damage (the large tanks are underground).

Another commenter wanted more discussion of plutonium storage. The DOE responded that the transuranic wastes likely have some plutonium, and that plutonium storage *per se* was not part of the EIS. They then pointed to more than a half dozen other EIS documents that did discuss it.

Several stakeholders wanted to know more about the CIF and the proposed vitrification facilities. The DOE provided some additional discussion, including the legal context and EPA's and South Carolina's roles. As part of this interchange, a suggestion was made to recover what otherwise would be waste energy from the incineration facility. The DOE responded that it was not economically effective to do so and also would require a permit from the EPA, which would be difficult to secure.

Other questions were posed about the adequacy of monitoring and plans for reducing or increasing the site labor force. The DOE provided straight-forward answers to these questions, which are consistent with what is presented in the EIS. In one case, the DOE noted that they inadvertently omitted an issue from the EIS; they then provided the data.

Compared with some other EISs, few of the remarks were made with much drama or responded to in that way. I was not at these meetings, so I cannot judge the expressions or body language. However, the transcript suggests intel-lectual exchanges rather than emotional confrontations.

## Changes in the plans proposed in the 1995 waste management EIS

Circumstances change and require that plans captured in EISs be alterable. Each federal department and agency has developed processes for altering its plans. In the case of the EIS and the DOE, it can issue a supplement or a new EIS. The DOE prepares what it calls a "supplement analysis" to determine if a new EIS is required or no further NEPA review is needed. Here I illustrate this process for the SRS waste management EIS.

In May 1997, the US Department of Energy (1997) decided to take several additional steps to manage mixed low-level radioactive and transuranic wastes that it said were consistent with the preferred alternative described in EIS-0217. One step was to send elemental mercury and other mercury-contami-nated low-level radioactive waste off-site for treatment. Residuals from the process were to be returned to SRS. Second, the supplement calls for vitrifying uranium chromium solutions and contaminated soils. And the supplement calls for building and operating two additional buildings to manage mixed and low-level radioactive waste and transuranic wastes. The five-page supplement assesses the impacts as small and states that any that would be found could be mitigated (*ibid.*).

This supplement is one of many actions across the DOE Complex that influ-ences this 1995 waste management EIS. Indeed, SRS publishes "What's New" on its home page (www.srs.gov) that lists, among other things, current NEPA actions that affect SRS. A recent interesting supplement serves as a second illustration. In the 1960s, the DOE produced plutonium containing oxide materials at the SRS site, so-called "low-assay plutonium" or LAP material. It was shipped to the Hanford site for scientific analyses but never opened. The supplement calls for shipping twelve drums containing approximately five kilograms of LAP material by truck (three drums per truck) back to SRS. The material there would be stored and then treated in the DWPF.

# Interview

Dr Kevin Brown is a research scientist in the Department of Civil and Environmental Engineering at Vanderbilt University, where he applies life-cycle and risk-analysis methods to radioactive waste management. I interviewed him on June 29, 2010. From 1985–2002, he worked at the Savannah River site for DuPont and Westinghouse on some of the more challenging radiological waste management problems in the world.

My questions focused on several of the projects described in the EIS and the overall concept behind a site-wide EIS. For context, he noted that SRS has no single-shell tanks *per se* and the SRS tanks are either double-shell or built in a teacup design, which means that leakage is into the teacup (annulus) and not into the environment. He could only think of hearing about one 10-gallon leak into the environment at SRS. This is in strong contrast to the Hanford site (Washington), which has a majority of single-shell tanks that do not have the teacup design, and hence has had a good deal of leakage to the environment. Furthermore, SRS had only two primary types of separation process and segregated high iron containing high-level nuclear waste from high-aluminum-containing wastes. This greatly simplifies the treatment process compared with a site such as Hanford, where there were many more separation processes and little attempt to segregate wastes.

Kevin Brown noted that the DWPF has been in operation for almost 15 years. (He was in the control room when it went hot.) About 10% of the waste volume goes to the DWPF, and this fraction accounts for about 55% of the curies and an even larger proportion of the long-lived radioactive elements. He feels that the operation has been smoother than he had anticipated. One major reason is that the materials retrieved from the tank have been more consistent than he anticipated, enhanced by the blending scheme that helps achieve a degree of consistency. This waste is then mixed with glass-forming frit and vitrified. The canisters of vitrified waste (see above for a description) are stored on-site in a special building. He praised the site for continuously monitoring the process, and observed that the production of the canisters was not optimized, because site personnel were concerned about changing a process that was working well and was not causing human health or environmental impacts.

The remaining 90% of the volume and about 45% of curies will be grouted (saltstone) and store permanently on site in concrete vaults. So far, the grout vaults have been effective in containing the waste. However, Kevin Brown suggested that 15–20 years is a short time to judge the ultimate success of this kind of waste management project and more information is needed. In fact, some of his current research focuses on the stability of the engineered system comprising the grout and vault. Overall, he believes that the EIS did a good job of capturing the important details of the DWPF and grout facilities.

The CIF was built in 1995 at a cost of $102 million and tested in 1997. The CIF could not efficiently incinerate the combustible wastes, which included radioactive elements. At an operating cost of $20 million per year, the site judged the facility not cost-effective for the limited feed material available, and instead site management found alternative management methods. In other words, the CIF was not used as had been anticipated in the EIS.

Transuranic wastes were another issue in this EIS. Transuranic waste has been going to WIPP in New Mexico. Indeed, the SRS Citizens Advisory Board congratulated the DOE on its 1000th shipment of transuranic wastes to WIPP. It also praised the DOE for examining options to treat more radioactive transuranic wastes on-site or ship them to WIPP.

The more general question is: What is the utility of this kind of site-wide EIS? Kevin Brown thinks this type of EIS is unusual; there are few that try to be so comprehensive. They are valuable to give a snapshot of management's thinking, but management have the option of changing their mind about individual projects, and circumstances change, which may mean that an EIS like this one does not even include all the plausible options for treating these wastes. In a sense, this kind of EIS is a site-wide scorecard that can be used to assess a comprehensive snapshot of site management thinking.

## Evaluation of the five questions

### Information

There is a massive amount of information in this document, so much that it is easy to lose track of its purpose. The focus, according to DOE, was the choice between no action and three levels of action regarding nuclear and some non-nuclear wastes. The choice the DOE made is risk-based, namely, to concentrate its resources on the most hazardous materials. It decided that this option would reduce worker exposures, compared with the more aggressive option, and manage the wastes more effectively compared with the least aggressive option.

Unfortunately, that message is buried in a sea of information. The quality and quantity of information brought to bear on the implications of this decision is impressive. Unfortunately, it is hard to access – often key numbers are in tables and/or graphs that are supposed to help explain the results. Sometimes the table and graphs are helpful, other times not. Some of the most important material is in even less scrutable appendices and graphs.

The most important shortcoming is the absence of a framework to support the choice of the chosen alternative. It would have been helpful to have had this decision grounded in basic risk-assessment and risk-management principles. The document addresses accidents, ecological impacts, and so on in encyclopedic detail. But it doesn't have an introductory section that describes the principles that it is silently evoking (Kaplan and Garrick 1981; Berlin and Stanton 1989):

- What negative impacts could result?

- What is their likelihood?
- What are their consequences?

Some sections contain the information to address these questions, for example, worker cancer risks (see above). But the discussion is so terse, and so unconnected to anything else about public or ecological health, that this reader had no idea how seriously this impact was compared with injuries from falls, accidents, and so on. Because the DOE's decision has consequences for risk prevention and cost, it merited an unequivocal effort to directly link risk events, risk likelihood, and consequences (Berlin and Stanton 1989).

More than any other project in this book, this EIS and this genre of EIS concern multiple projects. Because there is no framework, the reader is left to struggle with the relative significance of these different projects (DWPF, incinerator, trenches, vaults). It also seriously understates the importance of the maximum waste forecast, which needed to be unequivocally related to degree of site clean-up and future land-use options. The volume has too many dangling facts, and not enough integration and interpretation of the risk-related principles behind the choice of those facts. Overall, the tone and writing are acceptable; that is, the words are clear. But the message about the waste forecast and its relationship to key policy decisions is lacking.

On the technical side, I would have liked more discussion of the issue of uncertainty in modeling some of the impacts that were discussed, especially groundwater and air. However, this would be for the technically inclined.

## Comprehensiveness

This document is so comprehensive that it borders on the incomprehensible. There are more than two dozen waste management projects on this site, each attached to one or more other EISs. Decisions about future land use, degree of clean-up, and off-site decisions about transuranic and other waste forms add another layer of complexity. The authors tried hard to discuss everything, and in the process, as noted above, missed an opportunity to integrate using a widely recognized risk-analysis framework. I have no objections to reading an almanac or encyclopedia, but I do when the document is supposed to be a comprehensive summary supporting a key decision. Even if they could not, or chose not to, use an overall framework, they never made it clear what these projects had to do with each other. Kevin Brown, who is intimately familiar with the site, knows many of these connections. Having them in the report would have helped.

In other chapters, I have critiqued the economic and social analyses for being superficial. In this case, I found the economic and social impacts to be the most carefully presented in the document. A reader can understand how many jobs might be created by each of the different management options and waste amount options, how the management options were associated with jobs and income, and where workers would likely come from. In contrast, the water resources discussion (see the example above) and the accident scenarios were truly a challenge. I had to read each multiple times, prepare a table for myself to keep track of the numbers and assumptions, go to appendices, and then thumb back and forth because the discussions are separated from each other in the text.

## Coordination

The cooperating federal agencies were involved in the process, and their comments were not particularly challenging, leading this author to conclude that either the other agencies had had discussions with DOE before writing their comments and/or the EIS was mostly about issues that they had already addressed as part of their response to other EISs. That is, the key decision was for them perhaps a foregone conclusion.

## Accessibility to other stakeholders

Even in that pre-web-communication era, there were more than enough opportunities for public participation. However, the results were not what would be expected from facilities that handle highly toxic materials. The SSAB did not spend much time on this EIS (assuming that the minutes are a reflection of their concerns). Fewer than a dozen citizens testified at the public hearings. The limited amount of testimony is misleading because this document is much more of a recitation of multiple projects than it is a focus on any single decision. The SSAB had discussed nearly every one of these projects (e.g. DWPF, incinerator, groundwater), often many times. The public had heard about each of the individual projects on multiple occasions, often for many years. There was not much new for them in this EIS.

The "new" information is the choice of the mid-level treatment option and three waste levels. The only striking finding is the difference between the maximum waste impact and the other options. The report makes no attempt to hide the information. Indeed, the assumptions are described and illustrated with interesting and useful graphics. The maximum waste forecast does have implications for the future use discussions that the SSAB had been discussing, and yet I find no evidence of a strong reaction in the public comments. Perhaps this lack of reaction is due to the fact that even the maximum waste forecast uses only a small amount of land and confines almost all of the waste management actions to the central developed area. I found it fascinating that Tom Heenan of DOE mentioned this EIS on several occasions, and, assuming the minutes are a reflection of reality, there were no probing questions from the SSAB.

An EIS like this one, not built around a hot-button issue, nevertheless needs to be completed and is quite costly. Hence I believe that web-based and visualization technologies should be employed to make these kinds of EIS come to life for nongovernmental organizations, citizens, and the media. In Chapter 8, I formally suggest this option. Here, I illustrate it with a discussion of the DWPF. In 1982, President Carter was concerned about nuclear weapons-grade materials, and he moved the United States away from a leadership role in converting and reusing weapons-grade nuclear materials for fuel. The SRS DWPF is the most obvious implication of that decision. The SRS vitrification is a 42,000-square-foot, $2 billion-plus complex that blends aqueous nuclear waste with a borosilicate glass-forming compound, which then is poured into a stainless steel container to form massive canisters approximately 10 feet high, with a 2-foot diameter and weighing about 4000 pounds. These "logs" are stored on-site (US Department of Energy 1994).

The conversion of the waste material from liquid to a vitrified form is an important risk-based decision because it all but eliminates the possibility of the migration of the radioactive material. The DWPF has its own EISs. But it is mentioned multiple times in this one. The author can envision several drawings of the DWPF, the canisters it produces, their storage on the site, and maps that show where they are located relative to the concrete vaults where the less curie-laden radioactive waste is managed. A second set of graphical overlays could be made of the incinerator, and again its location shown on a map. Ultimately, a cumulative map would show the concentration of activities in the developed part of the site, and be accompanied by graphs and discussion that show the risk avoided by these choices. The cumulative maps of the three options and no-action option could be displayed against the backdrop of the associated risks.

## Fate without an EIS

I can't imagine that the DOE's decision was changed by this EIS. The DOE could not afford to cease work at the site because its legal agreements compel it to move forward and reach milestones. Furthermore, the region was economically distressed during this time, and the jobs and economic benefits are valued in this area. Yet there are likely more short-term health impacts on workers associated with the most aggressive waste management option. Furthermore, the cost (not reported) of the aggressive option likely was prohibitive.

# 7 Animas-La Plata, Four Corners: water rights and the Ute legacy

## Introduction

In 1868, after wars and skirmishes with the rapidly spreading Anglo-American population, the Ute Indian tribes moved to southern Colorado and northern New Mexico, the so-called Four Corners area, where Utah, Colorado, Arizona, and New Mexico meet (Figure 7.1). The Utes expected the new reservation to provide sufficient land and water. Their lands, however, were reduced and water was anything but plentiful. One hundred and fifty years after relocating to the Four Corners area, the Utes will receive a water supply from the Animas and La Plata rivers, the two major surface water sources, which settled their water claims.

This chapter has two objectives. The first is to illustrate an environmental assessment (EA) and Finding of No Significant Impact (FONSI) declaration, which were briefly described in Chapter 1. The reader will recall that an EA is a preliminary analysis. After completion of an EA, the federal agency must decide whether a full-blown EIS is required or whether it can declare a FONSI and move ahead with the project.

The second goal of the chapter is to describe one of the longest-standing political, moral, environmental, and social debates about a single project that has ever occurred in the United States; the debate involves multiple Congresses, Presidents of the United States and federal courts. The Animas-La Plata (ALP) project has had multiple EISs, supplementary EISs, and supplements to supplemental EISs. This chapter will attempt to tell the story of what has to be one of the most interesting applications of the NEPA process in US history.

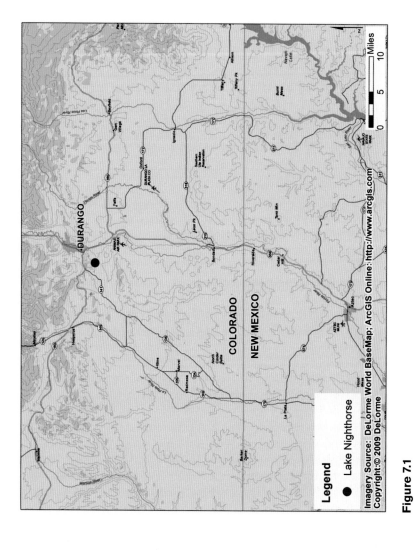

**Figure 7.1**

General Animas-La Plata project area showing location of Ridges Basin dam and reservoir

# Context

The history of the ALP water project begins with its geography. The Four Corners region is the only place in the United States where four states meet. Arizona, Colorado, New Mexico, and Utah share a common border demarcated by a brass plaque (Figure 7.1). The region is remarkably beautiful, with wooded mountain peaks reaching 10,000 feet and valleys of 4000 feet; and multiple national parks, such the Grand Canyon National Park (about 180 miles west of Four Corners in Arizona), Arches National Park (170 miles north in Utah), Bryce's Canyon Park (about 330 miles west in Utah), and Mesa Verde Park (63 miles east of Four Corners). In places, mineral deposits have been found, including gold.

While scenery and minerals are positive attributes, water on this high Colorado Plateau has been a major constraint. Average annual precipitation in the region is 12–15 inches, which means that the region is semi-arid. There is insufficient water for all the uses desired by residents, and water has proven to be an ongoing challenge.

After the 1848 US–Mexican war, the Four Corners area became part of the United States, and in 1868, after many skirmishes from Montana to the Mexican border, the Navajo and Ute tribes were relocated to reservations in the Four Corners area. As the value of the land became more obvious to Anglo settlers, the Indian tribes' territories were reduced and their limited water supply further constrained. With the main stems of the Colorado River to the west and the Arkansas River to the east, the two major surface water supplies in the area are the La Plata and Animas that flow through La Plata County in Colorado and San Juan County in New Mexico (Figure 7.1). In a region where there are four state governments, the Federal Bureau of Reclamation as a major landholder, county and local governments, and the Ute and Navajo nations, it is not surprising that the water rights have been a flash point for over a century.

In 2009, the population of the seven counties in the Four Corners area was 340,000, an increase of 6% since the year 2000 (US Bureau of the Census undated). The 35,000-square-mile area is sparsely populated with a gross population density of fewer than ten people per square mile, which is about the same as the states of North and South Dakota. A good part of the area has no population, and there are only two clusters: one in Farmington, New Mexico with a population of about 45,000; the second in Durango, Colorado with a population of about 14,000. Almost 120,000 of the people, about one-third of the total, self-identified as American Indian in the US Census.

## The populations: Utes, Americans, and environmentalists

### The Utes

The Utes were the first recorded population in the area we now call Colorado and Utah. They first met Anglos in the late sixteenth century, when they established relationships with the Spanish (Simmons 2000; Decker 2004; Rockwell 2006). After the 1848 war with Mexico, more Anglo-Americans moved west and decades of strife ensued, especially with the arrival of the Mormons. The tension grew rapidly after gold was discovered in the Pikes Peak area.

On March 2, 1868, the Utes signed a treaty that moved them to the Western Colorado territory. Notably, Abraham Lincoln had first asked the Mormons if they wanted some of the land; Brigham Young declined (McCool 1994). A series of treaty revisions took back most of the land the Utes received. Conflict, treaty revisions, and givebacks were the common pattern.

The Utes have had a difficult time trusting US leaders, with good reason. The Anglo-Americans wanted to convert these nomads to farmers, and at the same time were competing with them for resources, notably water. The Utes have a broader view of the significance of water than their Anglo counterparts, which persists in the twenty-first century. Dinar *et al.* (1995) report on a survey of Utes and Mormons, which asked about more than a dozen dimensions associated with water. Values exceeding 100 imply agreement between the Utes and the Mormons; values between 90 and 100 represent no relationship; and values less than 90 imply divergent views. With regard to practical uses of water, the number was 146, indicating strong agreement. But with regard to spiritual or religious significance attached to water, the number was only 10, and it was only 30 for the questions about the meaning of water.

In 1988, the Colorado Ute Indian Water Rights Settlement Act was signed by President Reagan. The act required that the ALP project water be provided to the tribes by January 1, 2000. If the water was not available, the Utes have the right to renegotiate. The Utes learned the hard way that good lawyers were essential to getting a fair shake of the resources. An article in the *High Country News*, which focused on ongoing litigation with the state of Utah over land and water, offers insight about Ute feelings:

> Elderly Ute woman: "you're like a rattlesnake on a hot rock. You've got the forked tongue."

> Luke Duncan, member of tribal council: "We feel like we just can't trust white people. The jurisdiction case is far from over; it's going to affect everything we do for a long time."

> Jonas Grant, tribal director of natural resources: "The people feel a lack of trust about the state; they keep trying to control what we have."

> While tribal council member Ron Wopsock added that there "will never be trust between the Indians and whites," a local Anglo was quoted as follows: "my biggest concern is being able to work together with the Indians. If we can't we're wasting our time."

> (McCool 1994, pp. 2–4)

## The Anglo-Americans

In 2006, when she was interviewed for an oral history, Stella Montoya lived on a cattle ranch in La Plata County, Colorado (Montoya 2006). Her family had raised sheep and cattle in Colorado and New Mexico longer than she could remember. Her father had been a sheep rancher, as was her husband for much of his life. She was one of five children and was the mother of six children. Yet only one of her children was a rancher, which she saw as an indication of the increasingly precarious position of ranching in this area with so little water.

At the time of this interview, she was chair of the water conservancy organization in the region, and her husband, who had passed away, had been chair for 30 years. Stella Montoya represents an extraordinarily well informed perspective on the importance of water in the region and relationships with the Utes and environmentalists, although her views are very personal.

Shortly after the interview began, she offered her first remarks about water:

Where my dad's farm was, we had plenty of water. But when we moved into La Plata, at first it wasn't so bad, but as time went on it got drier. They've . . . been working on that [La Plata] project for years, [the] early 1900s when they started working on that project. One time my husband came home and he was so happy because they were going to have it pumped through the mountains and come down. But then they took that project away from us. And every time, . . . they kept changing it. Every time they'd decide on something, then they'd move it because for some reason they couldn't have it there you know, the environmentalists kept moving it down, and so that's where it is now.

(*ibid.*, p. 4).

She described how her husband went to Washington, DC, and each time he returned they expected the project to get the final go-ahead, only to be disappointed. Studies would continue, but a final decision was not forthcoming. She noted that perhaps all they were doing was providing job security for Bureau of Reclamation employees: "I kept telling him [her husband], I said, that's just job security, that's all" (*ibid.*, p. 6).

She emphasized that the major hydrological problem is that when it rains, and when the snow melts, the water rushes down the mountains and is gone before it can be used. The dam is needed to capture and manage the flow. Rather than quitting the political endeavor to capture and store the water, she noted that they would get some water for irrigation because some of the water held back by the dam would leak into the ground and that water would be captured by the ranchers.

A key section describes her views of the Utes:

If it hadn't been for the Utes, I can tell you, we wouldn't have a project. Environmentalists tried their very best to separate the two of us, to make us enemies. (p. 8). The Utes said: We're partners and we're going to stay partners. (p. 9) We've worked as a team, Colorado people and our people and the Indians (p. 20).

Ms Montoya's opinions of the environmentalists reflected considerable frustration and anger.

Well some of the things they brought up, they were ridiculous (p. 16). They don't care about us. In fact, I think the squaw fish have more benefits. They talk about the endangered species, and we always figured out that we're the endangered species (p. 17).

Stella Montoya voices the all-too-familiar viewpoint that environmentalists are outsiders who try to impose their values on local people:

And see most of those people, they're not involved with ranching. And some, they would come from all over . . . would come to testify on these hearings and I don't think that's right. I think a person should be where they are because they know the locality and they know more about the country than somebody from Washington or California or anywhere else (p. 22). You know, people from back east, they don't understand our problems. Why would you want to build, spend so much money to build the dam to hold water? No it doesn't make sense to them (p. 24).

Among her final statement there was reference to the signing of agreements to build five projects on September 30, 1968. Her husband was at the ceremony, and while he got his picture taken with President Lyndon Johnson, she notes that their project (ALP) was the only one not built. "We're still waiting, hoping, working" (p. 25).

## The "environmentalists"

I admit to disliking the term "environmentalist" because it assumes values not always held. In the ALP case, the environmentalists were powerful opponents of the dam and lake, and they were remarkably effective in changing this project. Every impact element in Table 7.1 (see p. 184) was questioned over many years, and in multiple EISs and EAs. In essence, the environmental groups could not prevent the dam from being constructed. However, their arguments fundamentally changed the project. I cannot possibly do justice to the sophistication and breadth of their assertions. Hence I will focus on only one of their arguments.

Robert Wiygul, attorney for the Sierra Club, on behalf of the Earth Justice Legal Defense Fund, Sierra Club, Four Corners Action Coalition, and Taxpayers for the Animas River, submitted a set of arguments on about project economics (letter to Pat Schumacher, Bureau of Reclamation. April 17, 2000, www.angel fire.com/al/alpcentral/mycommentss.html). He began by asserting that the EIS failed to provide a benefit–cost analysis. Noting that the 1980 EIS contained a benefit–cost analysis, the Bureau, he asserted, had gone on record as acknowledging the relevance of such analysis to this project. In 1995, the Bureau updated the analysis. Having twice included benefit–cost analyses, he concluded that, in concurrence with NEPA requirements, a new update of the benefit–cost analysis should have been in this document. It was inappropriate to not update the benefit–cost analysis simply because the project is now being built to settle Indian water rights and claims.

Wiygul advanced the argument that NEPA requires consideration of factors beyond environmental quality, especially if those factors are relevant to decisions. This EIS, he argued, should have spelled out who will pay for the project, because the project had become more expensive. Next, he noted that the discount rate (interest rate charged for borrowing) had changed since the agreement was signed, and the higher cost meant that the benefits were only about half compared with what had been anticipated in the initial analysis.

An updated benefit–cost analysis, he concluded, would show that the project should not be built. Environmental groups offered alternative nonstructural options, such as redistributing water rights to the Utes and relying on diversion from other storage facilities in the region.

Wiygul continued by criticizing that the EIS did not specify what the water was going to be used for. Quoting directly from the purpose of the project in the supplemental EIS (SEIS; see immediately below), he asserted that the EIS was not legally acceptable because, without a specified use for the water, there was no possibility of developing competing alternative actions to compare with the preferred option. (This is grounds for rejecting an EIS for not meeting the NEPA no-action alternative requirement.)

Attorney Wiygul concluded that the only logical use for the water was that the tribes will market it. He argued that the tribes will not be able to easily market water because they will encounter jurisdictional issues with state governments and they will be responsible for operation and maintenance costs, which would be a serious problem for them. He concluded: "no private sector investor – whether for-profit or not-for-profit – would direct resources into such a fanciful scheme, regardless of beneficiaries" (letter, *op. cit.*, p. 3).

The economic arguments continued for many more pages. This sample is sufficient to demonstrate the sophistication and flavor of the environmental groups' arguments against a project that many of them believed was an environmentally degrading economic boondoggle.

## The Animas-La Plata saga

The purpose of the ALP project was summarized in the Final Supplemental Environmental Impact Statement (FSEIS):

Implement the Settlement act by providing the Ute Tribes an assured long-term water supply and water acquisition fund in order to satisfy the Tribe's senior water rights claims as quantified in the Settlement Act, and to provide for identified M&I [municipal and industrial] water needs in the Project area.

(*Federal Register*, January 4, 1999, section 1.3, p. 1.9)

The project certainly did not begin with those objectives. The official authorization was in 1968 as part of the Colorado River Basin Project Act (P.L. 90-537). The initial idea was for 191,200 acre-feet of water to be stored and used for irrigation, municipal, and industrial water uses in Colorado and New Mexico.

The law was passed prior to NEPA. The first sets of EISs were written in the 1970s and the FEIS published in 1980. However, President Carter ordered that no new water projects be started.

The abrupt stoppage likely would have ended this project without the Utes. In 1988, they reached a settlement with the federal government. The Colorado Ute Indian Water Rights Settlement Act (PL 100-585 as amended in 2000, PL 106-554) was passed, and it redirected the water project from primarily providing irrigation water for Anglo-Americans to the Utes and local users. However, additional changes were required when the Colorado pikeminnow, an endangered species, was found in the area.

In 1999, the Bureau of Reclamation examined not only structural alternatives, most notably the dam, but also so-called nonstructural options such as reservoir storage in different places, a cash outlay to buy water and bring it to the region – a total of ten alternative projects. Ultimately, the structural option prevailed. An agreement was reached in the year 2000 to scale it back to 120,000 acre-feet and to limit the flow out of the reservoir to 57,100 acre-feet (Draper 2006). Overall, 62% of the water is allocated to the tribes and 38% to nontribal bodies (Rodenbaugh 2009).

While the so-called ALP-Lite project moved ahead at a slow pace, the cost escalated to an estimated $500 million, prompting a year 2003 meeting by the Bureau of Reclamation to discuss the escalation of costs from $337.9 million to $500 million. The approved ALP plan included a pumping plant on the Animas River, and a 2.1-mile-long, 76-inch pipeline between the pumping station and a 276-feet high dam and reservoir (Lake Nighthorse, see Figure 7.2), which was named after retired US Senator Ben Nighthorse Campbell, a long-time advocate. The project also includes an outlet tunnel used to release water when required, and a pipeline for the Navajo nation from Farmington, New Mexico to Shiprock (about 22 miles, see Figure 7.1).

The construction of the pumping plant began in April 2003 and was completed in April 2004. In July 2006, the *Denver Post* (Draper 2006) reported that the project was 40% complete. Preliminary work for the dam construction also began in 2003, and a great deal of the work on the dam was completed in 2004. On October 17, 2008, *The Durango Herald* reported that the ALP project

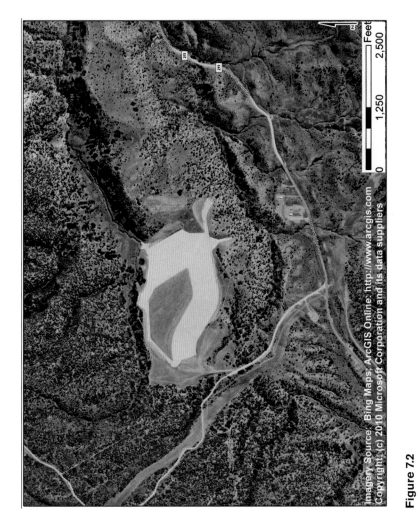

Imagery Source: Bing Maps; ArcGIS Online: http://www.arcgis.com
Copyright (c) 2010 Microsoft Corporation and its data suppliers

**Figure 7.2**

Lake Nighthorse

was 97% completed (Rodenbaugh 2008). The lake eventually was to cover about 1500 surface acres. The pumping station went online during the spring of 2009 (US Bureau of Reclamation 2009). Also in 2009, the US Fish and Wildlife Service (2009) noted that a fish hatchery located in western Colorado was going to introduce 100,000 rainbow trout into the Animas River at Durango in late June 2009. This was partly to address any losses from the dam project and from a disease that was affecting these fish.

In June 2009, $12.1 million of the American Recovery and Reinvestment Act was appropriated to ALP to be used to build a Leadership in Energy and Environmental Design (LEED)-certified building for the project and for part of the pipeline from Farmington to the Navajo Nation (US Bureau of Reclamation 2009). The US Department of the Interior (2010) annual report for the year 2009 discusses the project, requesting $27.4 million to build a pipeline from Farmington, New Mexico to Shiprock for municipal and industrial uses.

In June 2010, Robert Waldman, the Bureau's environmental specialist for the project, indicated that the reservoir was 63% filled (phone conversation with the author, June 30, 2010). The schedule for complete filling would be determined by snow melts and rainfall, and the area had been receiving below-normal rainfall. Nevertheless, he expected the reservoir to be filled by the end of 2011.

## The 2002 Final Environmental Assessment

On July 14, 2000, the US Bureau of Reclamation (2000a,b), in cooperation with the US Environmental Protection Agency (EPA) and the Colorado Ute Indian Tribes, released an FSEIS for the ALP Project. This FSEIS added to those of 1980, 1992, and 1996 (US Bureau of Reclamation (1980, 1992, 1996). Secretary of the Department of Interior Bruce Babbitt executed a Record of Decision on September 25, 2000 (US Bureau of Reclamation 2000b), which adopted an ALP project that would involve construction and operation of the system described above.

The year-2002 FEA (US Bureau of Reclamation 2002) in many ways was an afterthought, or perhaps better labeled a residual action required by the larger project. By the time it was written, the tough, multi-decade-long political decisions had been made. Nevertheless, three pipeline relocations had to be completed before the dam could be built and operated. Like other EAs, this one follows the form of an EIS, but with less detail. For example, the FEA has two technical appendices. However the main text is only fifty-seven pages. The EISs in this book were over 250 pages long and some were much longer. In order to present the flavor of the document, I have followed its sequence, which I have not necessarily done in the other chapters of this book.

The stated purpose is to relocate three pipelines so that the Ridges Basin Dam and Reservoir could be constructed. The report also includes a brief discussion of the modifications of a natural gas distribution pipeline, a county road, and electric transmission lines. However, these were not described in detail because the plans were not completed.

The three pipelines were as follows (see Figure 7.3):

- a 26-inch-diameter natural gas pipeline, owned by Northwest Pipeline Corporation
- a 16-inch-diameter natural gas liquids pipeline, owned by Mid-American Pipeline Corporation (MAPCO)
- a 10-inch-diameter natural gas pipeline, owned by MAPCO.

The project background section summarizes the entire ALP project, but quickly jumps to the pipeline relocations. What this means is that the reader who is interested in the history would need to do additional searching to learn more about the context for the pipeline relocations (US Bureau of Reclamation 1980, 1992, 1996, 2000a,b). Only one person complained in a letter about this inconvenience, and the Bureau argued back that it was not necessary to

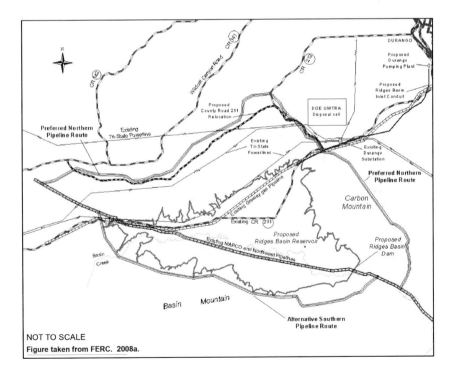

**Figure 7.3**

Proposed Northwest Pipeline relocation in relation to other Ridges Basin features

reproduce all the comments that already were published and available to interested parties.

Page 3 of the FEA identifies the need to relocate the pipelines, and mentions thirteen government agencies that had been contacted, including the Federal Energy Regulatory Commission (FERC), the US Fish and Wildlife Service (FWS), the Bureau of Indian Affairs and each Indian nation, and state and local governments. A full-blown EIS would have provided more details about the roles of each of these federal, tribal, state, and county agencies.

FERC is important as a cooperating agency because it is responsible for determining if natural gas facilities are in the public interest. FERC must issue a "Certificate of Public Convenience and Necessity" before a pipeline can be relocated (see Chapter 5). Northwest Pipeline Corporation, the pipeline owner, is responsible for obtaining the certificate from FERC. The FWS was pivotal because of concern about endangered species, especially golden eagle nests and elk calving areas, as well as other species described above, which led to a biological assessment. Consulting with the Bureau of Indian Affairs and individual Indian tribes was essential to avoid conflicts over disturbing culturally significant historical grounds (Winter *et al.* 1986). This reader infers that the Bureau of Alcohol, Tobacco, and Firearms issued a permit to use explosives, that the US Army Corps of Engineers issued stream-crossing permits, and that the US Department of Energy was consulted about the proximity of part of the pipeline to the DOE's uranium mill tailings waste storage cell near Durango, Colorado (the author has visited that site). However, an EA rarely goes into the same detail about organizational relationships as an EIS.

## Alternative actions

A no-action alternative is always required. In this case, however, the no-action alternative was not acceptable to the Bureau because the dam could not be built. Figure 7.3 shows that there are two basic options: one goes around the north side of the reservoir and the second around the south side. The Bureau examined seventeen alternative north and south options, finally settling on the two that are discussed in this FEA. Northwest and MAPCO proposed to build somewhere between 12.9 and 20.7 miles of new pipeline. They would abandon some of the old pipeline, and would not enlarge the pipeline capacity. MAPCO was considering changing its 10-inch line from gas to oil, which engendered considerable concern about a potential oil leak, leading to an oil risk report attached to the FEA. Except for a 0.6-mile segment, the design was to build all three pipelines in the adjacent 75-foot rights-of-way.

The report notes that, in addition to the 150-foot pipeline right-of-way, approximately 40–85 feet more would be required during construction. And small additional spaces would be needed for a metering station and other ancillary facilities. After contaminants were removed, some of the abandoned pipeline would be removed, and other parts excavated and left in place. Figure 7.3 shows the two major options.

The northern route is 6.9 miles long, in La Plata County, Colorado. The document points out that much of it is outside the Ridges Basin drainage area, and about half is owned by the Bureau of Reclamation, with the remainder owned by La Plata County. Much of the terrain is characterized as "low ridges with minimal visibility" (US Bureau of Reclamation 2002, p. 2-7). Unlike many other EISs and EAs, this one provides a feel for the terrain. For example:

> The last mile or so of the alignment at its west end skirts along the edge of an alluvial valley of Wildcat Creek before it crosses several ridges to reach the west tie-in point . . . In general, grades in or near valley bottoms and on ridge tops are relatively flat . . . One cross slope of about 26° occurs to the north of the HDD exit point, where potentially deep and/or unstable colluviums may exist.
>
> (*ibid.*, p. 2-8)

The southern route is 4.3 miles long and most of it is located within the Ridges Basin. The description foretells the decision against the shorter southern route: "The west end of the Basin Mountain is an identified landslide area" (*ibid.*, p. 2-9).

Regarding access to build the pipeline, the FEA describes the northern route as follows: "access to the proposed construction area is generally very good along most of the northern route alternative" (p. 2-10). In contrast, access to the southern route "is limited, with only two access points located along the alignment" (p. 2-12). Then the report once again identifies the southern route as steep, requiring "unconventional construction practices leading to a massive construction scar" (p. 2-12). The word "landslide" was used to describe the southern route in multiple places. The FEA continues that part of the southern route would require up to 275 feet of right-of-way in order to stabilize the construction in this difficult terrain.

The tabular data show that the northern route actually requires more land than its southern counterpart. However, this is because the northern route is longer. Before even describing the environmental impacts, the site descriptions clearly are making a case to reject the southern route.

The environmental impact section of the report examines a set of potential impacts, in essence concluding that the overwhelming majority are not significant and/or are temporary. Table 7.1 lists the impacts that were considered.

The disadvantages of the southern route continued to be emphasized. For example, "landslides are of concern for pipeline construction, primarily along the southern route" (*ibid.*, p. 3-2). The document points out that the northern route lies within 750 feet of the DOE Durango Uranium Mine Tailings Remedial Action (UMTRA) disposal cells for uranium mill tailing. However, the text adds that field inspections showed that the pipeline would not affect the "integrity of the UMTRA disposal cells site" (p. 3-3).

The very next paragraph once again sends a clear message about where this decision is heading. "Several geologically hazardous conditions were identified

**Table 7.1** Environmental impacts considered for Animas-La Plata, Four Corners

| Physical environment | Air quality |
|---|---|
| | Geology and soils |
| | Water resources and water quality |
| | Noise |
| Biological resources | Aquatic resources |
| | Wetlands |
| | Vegetation |
| | Wildlife |
| | Endangered and threatened species |
| Social and economic environment | Cultural resources |
| | Land use |
| | Transportation |
| | Visual resources |
| | Recreation |
| | Socioeconomics |

during review of the southern route, including landslides, construction on steep slopes, and rockfall hazards" (p. 3-3).

The discussion of water resources examines potential impacts on local groundwater as a result of converting the 10-inch natural gas pipeline to oil. For a project that was only proposed, a great deal of attention was devoted to this hypothetical. Frankly, the author was surprised to find a fourteen-page presentation about a petroleum products spill for a hypothetical project. I surmise that the Bureau of Reclamation would have been criticized for not addressing this possibility. But the Bureau, I believe, could have asserted that it would assess the risk and impact of a spill upon receiving a final proposal for a conversion of the gas line to oil. I was unable to determine why they chose to add this report to the document. However, after reading the document, I concluded that the Bureau was under pressure to avoid anything that could stop a project that had achieved enormous political support after decades of political and legal debate. Hence I speculate that the Bureau chose to be proactive about a hypothetical risk, rather than ignore it, which is not what the vast majority of EAs I have read have chosen to do; that is, they do not assess hypothetical impacts of hypothetical projects.

The oil spill analysis discusses why pipelines are used, the potential toxicity of a spill, and what the Bureau would do to minimize the risk. The document notes that pipelines are the safest way of transporting oil, much safer than tankers, ships, trucks, and railroads. That is, deaths, injuries, and spills are much more frequent with other transportation modes, which may also be much more expensive. Yet the report acknowledges that pipelines can fail and that the result can be injuries, ecological damage, and water pollution.

The oil spill analysis summarized US Department of Transportation data about the frequency and impact of oil spills. It describes various kinds of petroleum products, including crude oil, blended stocks, diesel fuel, fuel oil and gasoline, jet fuel, kerosene, toluene, xylene, benzene, and others. Benzene, which is about 2% of gasoline, is the fraction of greatest concern because of its toxicity at low concentrations, and the report uses rainbow trout as a sentinel at-risk organism.

The authors describe the worst-case scenario for the northern and southern alternatives:

> The worst possible case scenario for petroleum leak or spill in Ridges River Basin Reservoir would be during winter when low air temperatures slow evaporation of petroleum product components, such as benzene . . . Under this scenario, toxicity levels of benzene during a large spill (1677 gallons) would be the same as with the full volume of 120,000 acre-feet [0.002 mg/l in the first hour] . . . the lower volume of Ridges Basin Reservoir would still provide sufficient dilution to eliminate toxic affects of benzene.
>
> (*ibid.*, p. A-1)

The report then goes on to note that evaporation would occur, but that some heavy hydrocarbons would persist in the new reservoir and get into the food chain. The key observation is that only a small portion of the northern route drains into the reservoir, whereas almost all of the southern route drains into the reservoir:

> Although the northern route has a slightly greater risk associated with a pipeline leak, because of the greater length of pipeline, the risk of reaching Ridges Basin Reservoir is not as great as the southern route because of the more gentle terrain associated with the northern route . . . Because the southern route is located on steep slopes, the likelihood of a leak reaching Ridges Basin Reservoir is considered to be greater because of the steep gradient toward the reservoir. We expect [a leak] to move quickly, both below ground and above, if a leak occurred from a pipeline on the southern route.
>
> (*ibid.*, p. A-2)

If the oil pipeline change was made, the report notes that the Bureau would install block valves at either end of the pipeline to control the flow. The key conclusion, consistent with the FEA message, is that the northern route is preferred. The report then describes other steps to be taken to prevent a spill, and to reduce it should one occur.

The message about the northern versus southern route continues with wildlife. The FEA focuses on the golden eagle, elk and mule deer, again emphasizing the advantages of the northern route. Golden eagle nests are identified as sufficiently far away from the proposed northern route to not be a problem. The major concern is noise caused by blasting and visibility. The authors expect attenuation of noise impact by steep ridges. They noted that the closest wintering eagles were found along the river, 1–2 miles east of the northern route. The Bureau pledges to avoid construction within a quarter mile of an eagle's nest from December–June in the three areas along the northern route where they were identified.

With respect to elk, the report is clear about the preference for the northern route:

> Construction and operation along the southern route would significantly impact both elk calving areas at the base of Basin Mountain and elk migration over Basin Mountain. Impacts to elk calving areas from construction and operation along the northern route are not considered to be significant.
>
> (*ibid.*, p. 3-11)

In remote areas, endangered and threatened species are always a consideration. The Bureau of Reclamation prepared a biological assessment, which was attached. Their conclusion, arrived at jointly with the FWS, was that each of the federally listed species was unlikely to be adversely impacted. The biological assessment examines the potential impact on Colorado pikeminnow, razorback sucker, bald eagle, and southwestern willow flycatcher, as well as two federal candidate species. The two fish species of concern could potentially be impacted. Accordingly, two minor adjustments were made in the construction plan. One was not to use local water to test the pipeline; the second was to routinely monitor the pipeline with on-site inspections and remotely by technology. The bigger decision had already been made, which was to limit the average annual withdrawal to not more than 57,100 acre-feet of water and also to operate the reservoir to meet the needs of the endangered species (see above).

Unlike the oil spill analysis, the twenty-page biological assessment was not surprising, because there is nothing hypothetical about these potential impacts. The key point made in the biological assessment is that consultations were held between the Bureau of Reclamation and the FWS. Notably, this section ends with six bullet points that are aimed specifically at reducing the risk of a spill from the pipeline, and responding should a spill occur. In short, while the bulk of the appendix provides considerable detail about endangered species in the region, mostly in tabular form, the message is that the Bureau of Reclamation has a plan to reduce any risk. This appendix, along with the oil spill appendix, demonstrates that the Bureau is proactive and not afraid to deal with sensitive environmental issues.

Cultural resources of this area had already been surveyed extensively during the EIS processes for that dam. The pipeline relocation project studied a 500-foot-wide corridor along the pipeline route. The FEA commits to following an agreed-upon research protocol planned in consultation with the tribal nations, avoiding sites and recovering artifacts as required.

The visual impact again clearly distinguishes between the northern and southern routes. Construction of the northern route, the report notes, could result in "significant" visual impact during construction. The impact, however, is characterized as temporary, that is, for 3–5 years before revegetating occurs. Furthermore, the report stipulates that a landscaping plan will be developed that tries to minimize the visual impact. Several of these actions are identified. For example, "a directional drilling construction technique would be used to bore through Carbon Mountain, thereby reducing the potential for visual scarring to occur on Carbon Mountain as a result of the new pipeline alignment" (*ibid.*, p. 3-21) The analogous description for the southern route is less charitable: "Construction of the southern route alternative would result in temporary and permanent significant visual impact" (p. 3-21).

Other potential impact areas listed in Table 7.1 are summarized in a sentence or two as not significant or not new impacts; in other words, they had been in one of the earlier EISs. This includes even socioeconomic impacts. Clearly, some construction employment would be required, but it is not discussed as it is in many other EISs and EAs.

In addition to the three pipelines, the FEA briefly touched on three other projects because the Bureau of Reclamation interpreted them as having a common geography and timing, that is, related to the larger dam project. These are relocation of a county road, an electrical transmission line, and an 8-inch gas pipeline. The descriptions are brief, and the Bureau of Reclamation notes that it will use NEPA processes to undertake additional studies as required to conclude that the cumulative impact of these projects, and the three pipelines that are the focus of the FEA, have not changed from what already has been considered in the larger ALP project.

Environmental justice and Indian trust assets are addressed in two sentences, which in essence indicate that the Bureau of Reclamation has an agreement with the tribes that it will adhere to.

By the time the reader arrives at the fifth section, which is called "Recommended Action and Lists of Commitment", the recommendation is obvious. The northern route is preferred because of its access and less dangerous geological conditions, as well as the potentially important ecology, visual impact and potential erosion associated with the southern route. The cultural impact along the northern route is described as avoidable or capable of being mitigated by following the agreement signed with the tribal nations.

## FONSI declaration

The eight-page FONSI closely resemble dozens that I have read. I use it to illustrate the form of the declaration, as well as the substance of this specific case. It begins with a one-page letter from Carol DeAngelis, the area manager for the Bureau of Reclamation's Upper Colorado region to "interested agencies, Indian tribes, organizations and individuals" (FEA cover letter before FONSI). The letter mentions the three pipelines that were to be relocated and, the preferred northern alternative, and concludes that the "northern alternative will not result in any significant impact on the environment other than those previously identified in the Final Environmental Impact Statement . . ." (FEA cover letter before FONSI). A phone number is provided for anyone with questions about the environmental assessment or the entire project.

The FONSI declaration itself introduces the project as a whole and the need to relocate three pipelines, once again mentioning the choice of the northern route. It mentions the history of the project, beginning with 1980 and the Colorado Ute Settlement Act Amendments. The background also includes two paragraphs about the need to relocate the pipelines, and the Bureau of Reclamation's role as the lead, and then mentions the cooperating federal agencies such as FERC.

The next one and a half pages summarize the alternatives. The no-action alternative is presented in two sentences as a choice that would stop the entire project. After mentioning the consideration of seventeen alternatives, the declaration summarizes the southern and northern routes. The reasons for choosing the northern route are presented, including a potential oil spill from the hypothetical oil pipeline, easier access to the site, and space for temporary work along the northern route, as well as much more severe slopes and geological hazards along the southern route. The FONSI also expresses concern about possible impacts on calving grounds, erosion, and visual impacts along the southern route. The declaration notes potentially greater cultural impacts along the northern route, but asserts that they will be avoided or mitigated.

The next three pages present a snapshot of environmental impacts, including air quality, geology, soils, water resources, noise, vegetation, wildlife, endangered and threatened species, cultural resources, land use, transportation, visual resources, and recreation. The presentation consists of simple declarative statements with citations back to the FEA and a variety of EISs. Most of the statements are three or four sentences. The longest is about visual resources, and I quote it here as an illustration of the kind of messages being transmitted, I believe, to persuade readers to go along with the FONSI declaration:

> The FSEIS outlined several measures to be implemented by Reclamation to help reduce impacts associated with the construction and presence of the physical component of the ALP Project. The FEA analysis indicated that no new effects that previously identified with the FSEIS would occur. As indicated on page 3-283 and page 5-20 of the FSEIS, Reclamation would employ the services of a qualified landscape architect to develop and supervise implementation of a landscaping plan that specifically focuses on minimizing the visual impact of the pipeline relocation project. Measures specific to the pipeline construction include:
>
> 1. Areas graded and entrenched along the right-of-way would be restored to original grades.
> 2. A directional drilling construction technique would be used to bore through Carbon Mountain.

3. Contour slopes following backfilling pipeline trench to blend with existing terrain.
4. A visual mitigation plan would be developed for the corridor and would include measures to reduce the long-term visual impact of the right-of-way (requirement for the FERC approval process).

(FEA FONSI, p. 6)

The final two pages of the declaration describe coordination, beginning with a statement about the distribution of the draft FEA to 197 government and nongovernment parties – in essence, anyone who is interested. The fact that eleven comments were received is noted, as is their origin. The declaration points out that several commentators have called for a full-blown EIS, but the FONSI concluded that "it was determined within the FEA that since no new significant environmental impacts are associated with the abandonment and relocation activities, that the FEA would fully meet all the NEPA to compliance requirement for this action" (FEA, FONSI, p. 7).

The final section of the declaration describes consultation with twenty-six Native American tribes and the FWS. The conclusion is that no new concerns were identified. Hence the recommendation follows that "Reclamation, within the FEA, selected the Northern Route as the preferred alternative for relocation of the three gas pipelines. This Finding of No Significant Impact [underlined in FONSI] has determined that implementation of the preferred alternative will not have any significant effect to the human environment and that relocation construction activities should proceed" (FEA FONSI, p. 8).

This FONSI, like others I have read, is terse and reassuring. It is an effort to simply demonstrate that decision is justified by the facts and that the process followed regulations.

## Public reactions

By the time this FEA was written, the project was well under way and the vitriolic public and political debates were over. Nevertheless, following regulations, the Bureau of Reclamation held public hearings. A scoping meeting was held in Durango, Colorado in November 2001. Seven people commented on the pipeline relocation project. The public comment period was closed in January 2002, and the Bureau received fourteen letters, e-mails, and notes, which, according to the Bureau, were considered before the draft EA was prepared. The draft EA was released in April 2002: eleven comments were received from the public and from federal agencies.

Several members of the public testified or submitted notes. One, for example, thanked the Bureau for a chance to review the project and stated that she was in "full support of the northern route"(FEA, Appendix C, comment 1). A second was troubled by what he stated was plagiarism of parts of the oil spill analysis from another EIS. This writer added that the EIS was in the public domain, and hence the quotation was not illegal. However, not to cite the authors "borders upon unethical" (FEA, Appendix C, comment 6). The Bureau responded that this oversight would be corrected.

One writer was concerned about impacts on golden eagles, and urged the Bureau to work with the FWS to protect endangered species. The Bureau agreed. This writer was also concerned about the potential impact of an oil spill from the rebuilt MAPCO pipeline. The Bureau responded that if MAPCO does follow through, then further analysis will be undertaken.

FERC provided several comments. One was a concern that the analysis of the relocation of County Road 211, the Greeley Gas company facility, and the Tri State electrical transmission facilities were not presented in sufficient detail in the document. The Bureau's answer was that none of these facilities had sufficiently detailed plans to warrant action at that time.

A local county representative pointed out that they were considering options for realigning Route 211 and would notify the Bureau of Reclamation when final decisions were made. A staff member of the FWS indicated he had spoken with a Bureau staff member and had nothing else to add. A regional manager of the DOE offered three comments that required minor revisions of the text. A representative of the State of Colorado indicated that the issues they had previously raised had been addressed, with two exceptions, which were relatively minor.

The remaining comments were more interesting. Representing a not-for-profit organization, Steve Cone and Vena Forbes Wilson argued that a full-blown EIS was necessary to address a possible pipeline rupture, and that the cumulative effects presented in the study had not been sufficiently addressed. The Bureau responded that the oil spill analysis presented in the FEA would be updated if MAPCO decided to move forward with converting their gas line to oil, as would the cumulative effects analysis. In June 2002, MAPCO sold the line in question to another company, and there has been no proposal to convert the line to oil (Waldman, phone conversation, 2010, *op. cit.*).

Douglas Grew, a local resident who lives near and drives by the northern half, opposed the choice of the northern route. He feared bisecting the animal migration path. The agency responded that this had been extensively studied and that it believed the impact would be limited, even during active construction. Grew added that more cultural resources would be disturbed by the pipeline on the northern route than on the southern side. The Bureau disagreed, asserting that the sites would be avoided if at all possible, and if they could not be avoided, then the cultural artifacts would be removed. Grew asserted that more recreational sites and forest land would be disturbed, and more would have to be spent, than on the southern option, and that access would be difficult on the northern slope route. The Bureau commented that access was even more difficult on the southern side and that overall the greater potential impacts were clearly on the southern route.

Grew continued that the scar created during construction was intolerable to many, and that it was "clear that those of us who both live along Wildcat

Canyon Road and drive on it on a daily basis do not count much in this analysis" (FEA, Appendix C, comment 4). The Bureau's response was that visual impact would be reduced by locating the pipeline in less visible and less prominent areas, and that a scar on the south side would be even more visible.

Representing the Cedar Hill Clean Water Coalition, Jacob Hottell argued that all bald eagles might be harmed by the proposed relocation, and that the project would reduce the flow of the Animas River. The Bureau did not agree with the second point, and responded that the biological assessment in this FEA report, along with the FWS, agreed that bald eagles would not be threatened.

Perhaps the most interesting comment was from an attorney representing the Pueblo of Laguna. The letter did not object to the proposed action. Rather, it was a statement of the Pueblo's understanding of their agreement with the Bureau of Reclamation about the project. The Bureau did not contest any of the assertions. As noted earlier, the tribal nations had learned the need to be represented by attorneys in their dealings with the Anglos.

## Interview

Dr Frank Popper is Professor of Planning and Public Policy at Rutgers University, and also teaches at Princeton. Frank is well known for originating the acronym LULU to describe locally unwanted land uses, such as landfills, incinerators, and factories. In 1987, Frank and his wife Deborah wrote a short paper, "The Great Plains: from dust to dust" (Popper and Popper 1987), which asserted that the semi-arid Great Plains of the United States, Canada, and Mexico were losing population and jobs, and that it made sense for the region to become a "Buffalo Commons" of perennial grasslands where the buffalo and other native species could roam. The Poppers received considerable hate mail from the region accusing them of being easterners with little knowledge or stake in their words. Yet the Poppers' ideas have become increasingly accepted and embraced in the Great Plains.

I asked Frank Popper some questions about the ALP project because I felt that he could place ALP in the context of changes that have been occurring in the west. I interviewed him on June 30, 2010. He noted that Animas-LaPlata is:

actually a hangover from a generation ago. In 1986 the Bureau of Reclamation decided that it would no longer be a dam-building agency, but instead a water-conservation one. In practice this meant that it would start no new projects, and it hasn't. But it would try to complete ones existing in 1986, like ALP. As a result the Bureau is often blamed for practices it no longer endorses, as HUD [Housing and Urban Development] is blamed for high-rise low-income housing, which it has been legally prohibited from building since 1968. Animas-La Plata, whatever its fate, will be the last such large project in western Colorado.

Dr Popper noted that the Buffalo Commons idea was partly a reaction to these massive building projects, such as the 1944 Pick–Sloan Plan that created six large artificial lakes just east of and in the Great Plains by damming the Missouri River. The lakes, known as the Great Lakes of the Missouri, represented a compromise between the Bureau of Reclamation and the US Army Corps of Engineers.

He noted that Buffalo Commons, in contrast to these mega-scale projects, is a land-use vision for a sustainable environmental end-state. Frank Popper: "It is Plan B taking over from a Plan A that has been failing for over a century." He continued that the ALP project would not recur today for good reasons. Agriculture there is not competitive with agriculture elsewhere in the country and Canada, especially after the North American Free Trade Agreement (NAFTA). Raising cattle is the last choice, not a particularly good one. Animas-La Plata is an economic boondoggle that may or may not help local Indians. He noted that other similar projects are often not used as anticipated. Many dam projects, for instance, end up being used by weekend boat enthusiasts rather than for their ostensible agricultural purposes.

Popper sided with environmentalists in most cases because projects hurt ecosystems, and/or were economically inefficient. He admits frustration when politics trumps good EIS and other science-based analyses. He suspects that the BP Gulf of Mexico oil spill will change people's minds about large-scale projects in difficult environments. At a minimum, the disaster will lead to better analyses in EISs and risk analysis that decision-makers will take into account. He points to difficulties of dealing with methane in places like Wyoming, where natural gas and oil exploitation are taking place, as another current example of large-scale projects with poorly understood impacts. He said "too many games have been played with [environmental impact and cost–benefit] analyses over the years" by decision-makers who use these tools to get the decisions they want, rather than to understand their decisions' implications.

## Evaluation of the five questions

### Information

There is a limited amount of information in this FEA, and it is pretty clear how it was to be used to support the northern route. There is no plausible no-action alternative, and the southern route is portrayed as a bad choice because of the risk of landslides, impacts on elk calving grounds, a visible permanent scar, and a potential oil spill from a non-existent oil line into the Nighthorse reservoir. The limitations of the northern route are not avoided. But they are characterized as avoidable or capable of being mitigated, buttressed by some thinking about mitigation and resilience options. The pro-north route message is blatant and helped by writing that is above average for an EA and EIS. I could be mistaken; however, this EA feels slanted toward the northern route. Some language is devoted to mitigating problems that would be encountered; but not much.

My routine complaint is the lack of economic analysis. I do not understand why we are not told how many workers would be employed to build the pipeline, unless none of them were being hired locally, or the numbers are trivial compared with the numbers brought in to build the dam, reservoir, pumping station, and long pipelines to carry the water. Of course, had this information been included, I would not have needed to speculate.

## Comprehensiveness

The EA lists and briefly summarizes the impacts, with a few exceptions. The main focus is on landslides, access to the sites, visual impact, and others just noted. In other chapters, I have critiqued the economic analyses for being superficial. Here, there are none. Surely the longer northern route would, I assume, create more construction jobs and more injuries to construction workers. Perhaps I am wrong. But without the data, I was forced to guess.

## Coordination

The Bureau of Reclamation tried very hard to coordinate with the FWS because of the endangered species issues, and in general it methodically listed its outreach efforts with cooperating agencies. With the exceptions of FWS and FERC, these were not emphasized in the FEA. However, as noted earlier, I speculated that the oil spill and biological opinion were the product of the agencies trying to avoid any charges of insensitivity to environmental and cultural issues.

## Accessibility to other stakeholders

Powerful local and national stakeholders played key roles in this project. The EA does not do justice to the struggle between the Anglo-Americans, Indian Americans, and environmentalists, who had fundamental differences about the value and meaning of this project. For the Anglo-Americans, it started out as a chance to continue ranching as a well as to add more urban development. For the Utes and Navajos, it was an opportunity to get back part of what they thought had been promised to them in 1868. For the environmentalists, it was a chance to take an ethical position against large-scale projects with high cost–benefit ratios that also radically alter ecosystems. Each had multiple opportunities to argue in this EIS venue, before Congress, and in the courts. And for the Bureau of Reclamation, it was a project that they were carrying for all three groups, but especially the tribal nations. This project, more so than any other in this book, had the elements of a soap opera played out in multiple venues among parties with a lot at stake.

Technically, the accessibility of this EA could have been improved by using mapping technologies that would have allowed a simulated flight over the north and south routes to show the locations of landslide-prone areas, areas that would have been scarred by laying the pipeline, and many of the most visually oriented impacts discussed in this EA.

## Fate without an EIS

The EISs and EAs testify to a markedly changed project over two decades. The 1980 and 2000 EISs are different. Did the EIS contribute to this change, or were they the places to record the changes? I think more of the latter than the former. In the end, regional and national politics won out over science. However, having the EIS requirement and later other major pieces of action forcing environmental legislation made it possible for the Utes and environmentalists to press home their views and arguments. The tribal nations and the environmental groups smartly used NEPA and other tools to challenge the original idea, which primarily was water for Anglo-Americans.

Even the EA in this chapter, I speculate, was so carefully composed because the Bureau knew the opponents of the dam were looking for a legally actionable gap in the document to challenge the project. The well-conceived arguments by Wiygul demonstrate that in the year 2000, the environmentalists were still trying to defeat the dam, and in 2002, the tribes, in their comments about the EA, were trying to make sure that what they had agreed to was the Bureau's policy. Furthermore, a more practical reality was that the pipelines were buried beneath the dam location. They had to be removed because, as Waldman (phone conversation, 2010, *op. cit.*) noted, the Bureau's experience with removing pipelines underneath dams after the fact shows that risk and cost would be substantially increased. This was a project that was too far along to stop. It would have been larger and more destructive without the EIS process.

# 8 NEPA and the challenges of the early twenty-first century

....................................................................

## Introduction

The early twenty-first century has begun with a set of perplexing and complicated environment-related problems, and yet many twentieth-century legacy problems remain. In this chapter, I offer my views about improving NEPA, and suggest how it can be applied to several of these challenges. As a multi-decade reader of the *Environmental Impact Assessment Review*; *Risk Analysis: An International Journal*; *Environmental Planning and Management*; *Environmental Management*, and other environmental and public health analysis journals, I have encountered hundreds of ideas, suggestions, and recommendations about how to protect the environment and public health. Hence I make no claim that I originated these suggestions.

This chapter asserts that NEPA is the national equivalent of a land-use planning mandate and mechanism, and as such its strengths are gathering and analyzing information relevant to decision-makers, widely communicating that information through multiple pathways to multiple stakeholder groups, and offering flexibility to decision-makers. NEPA is not a substitute for directive legislation and regulations to achieve specific environmental policy objectives, such as clean water, clean air, and a sustainable environment. NEPA can be part of the federal government policy response, but only a part of it.

NEPA, like many other American environmental statutes, appears to be trapped in uncommon partisan politics, on a path that accepts small patches rather than the kind of comprehensive adjustment that would benefit from a massive infusion of science-driven analysis. Hence I have suggested some patches and only a few major revisions. In short, I have focused on better implementation (Pressman and Wildavsky 1973; Bardach 1977) of what are the law's strengths.

## NEPA and the EIS as the federal government's regional planning process

Among the large developed nations, the United States is unusual because it does not have a national planning law (Cullingsworth 1993). Speaking about the United States, Popper (1992, p. 48) said that "to gain support and eventual passage, most regional planning legislation must travel under the cover of clean air or clean water management." In 1974, the US Congress almost passed a national land-use policy act that had been introduced by Senator Henry Jackson between 1968 and 1975. It may be time for the United States to end its, I believe, counterproductive opposition to national and regional planning. But the history of local land-use control endures, and so, with the exception of some states that have tried state land-use planning, NEPA and the EIS process are the closest the United States has to a real regional planning mandate and process.

Looked at as part of a planning process, the EIS has many of the same strengths and weaknesses as city planning processes. City planners create site-specific and city-wide objectives. They develop alternative ideas for meeting those objectives, gather information about them, and map the alternative so their geographic expression is clear, and they evaluate the strengths and weaknesses of the options. The alternatives are rolled out at meetings, and are made available on websites, to the media, and to various others for consideration. The planning department takes charge, but other departments cooperate. The options are likely to be revised and viewed at multiple meetings, but in the end, with considerable input from powerful interest groups, the mayor, county freeholders, or whoever is in charge makes a decision. Those decisions favor some parties and disadvantage others. The decision-maker can be challenged in court, but rarely successfully. The city planning process is chameleon-like because it has to accommodate on-the-ground realties and changes, as well as considerable political input into the process.

The EIS process, like the city planning one, is imperfect, with nearly all of the same flaws. However, it shares the same potential strengths: gathering and analysis of information, multifaceted communication mechanisms, and the provision of flexible alternatives.

### Information

I am not inclined to nitpick information deficiencies, because EISs already have more information that the vast majority of us can absorb. I do, however, have problems with three kinds of information: regional economic, local environmental justice/socioeconomic, and cumulative impacts.

## Regional economic impact – information patch

The EIS for the Savannah River nuclear waste management site (Chapter 6), is the best socioeconomic analysis described in this book. It provides estimates of employment during construction and during the operation of the DOE's waste management facilities, and then it estimates if new employees will be needed from outside the region. It also provides data about the distribution of poor racial minorities.

However, although the best in this book, it is not even minimally acceptable. Every EIS should provide the direct impacts on jobs, regional income, and taxes. They do not. If these are significant, then indirect impacts should be provided. These are harder to estimate, but standard multipliers are available from the Bureau of Economic Analysis and a variety of economic models can be used when the activity starts to approach a billion dollars (Greenberg *et al.* 2007).

I realize that an EIS is not an *economic* impact statement, but it is supposed to consider the set of factors that are important to decision-making, and economic benefits and costs are always important. If EIS authors want to pretend that economic benefits are not important, they are providing decision-makers with a good reason to ignore the EIS.

## Local environmental justice and socioeconomic – information patch

The treatment of environmental justice is my second concern. The EIS described in Chapter 4, which is about a liquefied natural gas complex, was particularly troubling because of the efforts that seem to have been devoted to negating this as an issue. Maybe the environmental justice impact is trivial, but I did not draw that conclusion. The presentation covers multiple impacts. These include economic and environmental justice impacts associated with constructing and operating the LNG terminal in an area that is both disproportionately poor and African American. It also considers the economic impact of closing a major bridge due to LNG tanker passage; possible accidents; impacts of the facility on property values and insurance; demands on local police and fire departments; costs of providing security; and job- and income-creation. Data are presented to show that the LNG facility is compatible with existing land uses, but the report is too terse to be credible and there is no effort to integrate it into a coherent story.

Galisteo Consulting Group (2002) prepared a review of social, cultural, and economic impact assessments for the EPA, observing that this impact criterion lacks methodological rigor and sufficient guidance. Some of these EISs offer little to nothing about important elements. The LNG case ended by declaring all of them insignificant. And I cannot say why.

In the EIS described in Chapter 7, for Animas-La Plata, the arrangement with the Indian tribes is summarized in a few terse sentences. No effort is made even to describe how many Indian Americans live in the region, although the EA was preceded by multiple EISs. The light rail system discussed in Chapter 2 is built in an area in the United States that is heavily African and Latino American. The Ellis Island project (Chapter 3) is in that same area. Basic information about the population's socioeconomic and racial/ethnic background is present, but is minimal. There really is no coherent story told. More attention is paid to the number of fish in the Ellis Island area than is directly devoted to the disadvantaged populations likely to be visiting the site. At the Johnston Island site used to destroy chemical weapons (Chapter 5), several people testified to their views on transporting chemical hazards halfway around the globe from a Caucasian to an Asian area. No data were provided in the analysis itself.

Elsewhere, Greenberg and Cidon (1997) have written that environmental justice is a two-sided coin. Almost all the literature measures the economic, environmental, and public health burden on poor and minority communities. It should also measure positive attributes or lack thereof in these communities, such as playgrounds, good schools, job creation, and other facilities that appeal to people. I believe these attributes would speak to the benefits that some of these projects could bring to these regions and to the disadvantaged populations, if they were tracked.

Overall, after reading these EISs and EAs, I am persuaded that analysts do not know what to prepare and how to interpret the data, other than to dichotomize the impacts as significant or not. They prepare what appears to be a safe report. While this approach may be sufficient to clear regulatory requirements, it troubles residents, who can get more information on their computers and create their own maps and draw their own interpretations. In this regard, *Wikimapia* and *OpenStreetMap* (Goodchild 2007) are new computer-based tools that are rapidly increasing the capacity of interested people to obtain, analyze, and potentially contradict agency data.

Agency guidance, especially from the EPA, needs revisiting in the light of increasing public capacity. Unless EISs and EAs deal more effectively with the environmental justice and socioeconomic categories, there will be many more embarrassing public meetings for government agencies, who will find that local populations have assembled much more data, and much more accurate data, and that they are able to display it in much more useful ways than the government and their consultants.

## Cumulative impacts information – major addition

My third information-related concern is with regard to the cumulative effects. The problems are similar to those for socioeconomic status. EPA guidance (US EPA 1999) speaks to the need to measure the effects of projects that could impact resources (water, air, fish, forests, wetlands, etc.) and those that could involve several actions in the same area both in the present and in the future. For example, a poorly planned marina could lead directly to a loss of wetlands, disruption of wildlife habitat, and an increase in impervious surfaces. These direct impacts, in turn, would likely lead to additional stormwater runoff that would include sediment and contaminants, some of which would be harmful to fish. The long-term effects would be decline in water quality, creation of contaminated sediment hotspots, and likely a decrease in fishable and swimmable waters. These indirect and cumulative impacts pose a challenge to analysts with regard to defining the spatial and temporal scope of their responsibility. There is a challenging issue of increasing uncertainty as analysts try to translate direct impacts into indirect ones that may be located miles from the site and not occur until years into the future. The EPA guidance, accordingly, indicates that human impact assessments will not be present in every case.

This guidance, I believe, has been a strategic mistake. The EISs in this book, with rare exceptions, do not stray very far from the specific site or into anything other than the immediate future. The best example in the book is the LNG facility (Chapter 4). The EIS examines seventeen other public and private projects and thirteen dredge projects along the river system. Cumulative impacts considered include geology, soil, air quality, and so on. The conclusion regarding each impact is that the effects are insignificant. While I do not agree with their conclusion, at least an effort was made that allows the reader to agree or disagree. In contrast, much of what passes for cumulative impact analysis in EISs and EAs consists of two-sentence summaries of what has already been concluded for each variable.

The Ellis Island EIS (Chapter 3) was a missed opportunity to examine cumulative impacts. The phrase "cumulative impact" is found over 100 times. But the authors focus on the cumulative impact of every outcome variable, such as parking, noise, and so on. The real opportunity here was to tell the story of how Ellis Island was going to fit together with Liberty Island (Statue of Liberty), Governors Island, and more than a dozen other recreational projects in the harbor, the Hudson River, and adjacent areas in New York and New Jersey. In fact, the document offers one paragraph about all of these projects. The authors missed an opportunity to present the big picture – that is, the cumulative positive *regional* impact. There has been some tension between New Jersey, New York City, and New York State over who controls the land and revenue (Chapter 3). However, this should not have prevented the authors of this document from telling an important cumulative impact story across a multi-state region that could have a positive influence on the region for decades. It is disappointing that the opportunity to present the Ellis Island project as part of a grander plan was missed.

A distressing side of cumulative impact assessment is environmental justice. The tenth hazardous waste facility or the fifth highway adjustment in a local area may add little to measurable public health or environmental impact on an area. But the public likely will not believe that, and there are instances where the public has reacted with an outcry of environmental injustice, and sued. Specifically, in 1997, a group of residents in Chester, Pennsylvania claimed environmental discrimination under Title VI of the Civil Rights Act of 1964 and sued the Pennsylvania State Department of Environmental Protection, which had granted a permit to build yet another waste management facility in an area that already had multiple such facilities. The District Court dismissed the case, but the Third Circuit court restored it, ruling that private individuals could bring discrimination claims to enforce Title VI. This case was the first instance of a citizens group being granted such standing. Similar instances have occurred since that time, and the US Supreme Court has become involved ruling that Title VI did not create a private right of action to enforce these regulations. I do not want to spend any more time here presenting the legal arguments because they are well-documented and readily available on the web.

The issue for me was the message that was sent to residents of Chester by the unwillingness of their State Department of Environmental Protection to seriously consider that continuing to permit a cluster of locally unwanted land uses was an unacceptable cumulative impact. Comprehensive cumulative risk assessment and health impact assessment would be an aid to decision-makers willing to consider how equity, as well as effectiveness and efficiency, can be built into a plan (Sexton and Linder 2010). The arguments advanced in the LNG and Johnston Island EISs could have been brought out into the open with these tools, rather than dismissed as unscientific.

## Communications

Between 1966 and 1981, a set of laws such as the Freedom of Information Act, the National Forest Management Act, NEPA, and others called for disclosure and transparency. The communications effort behind the scoping document for the Hudson–Bergen rail line is an illustration of what is now common in NEPA. I am not implying that the internet directly leads to better decisions; but I am implying that the effort to reach people has increased. Below I offer some suggestions for enhancing information flow for what already should be a strength of the EIS process.

## Providing translations – patch

Richard Nixon was President when NEPA was signed, and the overwhelming majority of US residents were Caucasian and spoke English. Forty years later, the population has increased to over 300 million, and the population that the US Bureau of the Census describes as "Hispanic" has grown from fewer than 10 million to almost 50 million, and could exceed 100 million by the year 2050 (US Bureau of the Census undated). While many speak English, the United States has to deal with the reality that many residents have little understanding of English. Given the propensity for large locally unwanted land uses (LULUs) to be constructed in areas disproportionately inhabited by disadvantaged people, the reality is that notices, minutes, and EISs in their entirety, or at least summaries, need to be translated into Spanish and, in some instances, other languages.

The first major example of accommodating Spanish-language populations was Denver's I-70 East corridor. The EIS focused on rapid transit options between downtown Denver and the Denver International Airport (US Department of Transportation 2008). The EIS reported that 32% of Denver's population is Latino, varying from 17% to 83% along the corridor. Thirty-five percent of the population along the corridor spoke a language other than English (*ibid.*, 5.2-10). The public notices, summaries, and EISs were in Spanish, as well as English. Also, translators were at public meetings. M. Gonzalez-Estay, who was involved in the EIS, feels that the outreach was successful and will become common in areas where a large non-English-language population is resident in the impacted areas (personal conversation with the author, July 16, 2010).

## Electronic formats – patch

In 1970 there was no internet, and computers were for the select few who had punch-cards and access to large mainframes. In 2010, 76% of the US population uses the internet (Miniwatts Marketing Group 2010). EISs are now posted on the web, which makes an enormous difference to those who want to read them.

In addition, the younger population has become agile with visualizations and nonverbal forms of communication. In a democracy, if we truly want many people to understand the essence of what EISs are saying, it is prudent to develop visual aids to attach to them. A case in hand from this book is Animas-La Plata. I was able to find some photos of the area, but really wanted to see a flyover of the north versus south routes that would have allowed me to see the landslide-prone areas, the eagle areas, and so on. I imagined myself in an IMAX theater, seeing the alternatives come to life. I then re-read every case study in the book, and was able to envision similar graphical/visual aids. I realize that visual aids could substantially increase the cost. But, over time, it is likely that costs will go down. Furthermore, when you are contemplating spending hundreds of millions to billions of dollars on major projects, it makes sense to develop visualizations for those with limited time or capacity to read massive documents. Visualizations would allow people at least be able to get a sense of the project and its potential impacts. I realize that visualization can leave misleading impressions. However, I assume that a small set of readers are highly motivated and technically capable individuals who will still read the entire document.

While the literature on plan implementation is based on limited numbers of case studies, it does suggest that a wider variety of audiences improves planning outcomes. Hence visualizations, like translations, will increase the likelihood of a larger and more interested audience.

The US Department of Transportation has been experimenting with visualizations. An initial report (Volpe Center 2010) examined applications in Arizona, Ohio, and Wyoming. The designers compared before-and-after project views using aerial photos, single photos taken in the field, terrain models, standard plans and drawings, photo simulations, and animations. The report notes that the public was better able to see how the project would change the areas. The visualizations added 0.5–2% to the cost of preparing documents. Officials, the report notes, said that the visualizations were worth the investment because they helped people, even opponents, better understand the impacts, and ultimately saved time and money.

## Flexibility – patch or major addition

With regard to policy flexibility, NEPA, at least on the surface, has built-in flexibility: a no-action alternative and other options. But as we have seen in the case studies, many of these are non-starters. However, the process at least requires options.

NEPA regulations require consideration of a set of categories of environmental impact, and guidance exists on how these are supposed to be addressed in an EA and EIS. But with regard to the preparation of an EIS, each assessment borrows from standards, guidelines, and requirements from other federal and state laws, but allows analysts to focus on some more than others and to interpret the data in a regional context. This means there is considerable flexibility built into the process. This flexibility has had the disadvantage of allowing decision-makers to avoid focusing on key impacts; but for other projects, it has allowed them to zero in and carefully present their story.

With regard to enhancing policy flexibility, I consider three types of flexibility, which I will refer to as macro, meso, and micro. By macro, I mean stepping back from the project proposal and potentially changing it to be more acceptable to the parties. The EIS has to include a no-action alternative. But, as this book demonstrates, these are rarely feasible options because they contradict the agency's overall mission, or at least are so characterized.

A better option, at least in some cases, is to change the project, or some major elements of it, thereby allowing all the parties to claim a victory. In fact, this is standard practice in city planning. Mega-scale housing and office building development gets scaled back to projects that the parties reluctantly agree to. Animus-La Plata is the best example of macro-scale flexibility in this book. The original project was rejected and bitterly contested. The final solution gave the Indian tribes some water; it gave the Anglo-Americans a sense that something had been accomplished after decades of failure, including the ability to buy water; and it gave environmentalists a moral victory by scaling down the project and, at least theoretically, protecting endangered species.

Not everyone should be expected to support macro flexibility, because too much flexibility could lead to actions that have nothing to do with the agency and the original goals of the project. For example, Ellison (1999) criticized the Animas-La Plata compromise as not achieving the objectives of federal water programs and for being co-opted by interest groups (Indian tribes, local Anglo-Americans, environmentalists), including the Bureau of Reclamation, to justify an unjustifiably expensive project. I do not agree. The Animas-La Plata project ultimately was supported by a wide range of interest groups. It tried to accommodate ecology, hydrology, and other environmental considerations, and address the Utes' long-standing water claims. I do not doubt that the final benefit–cost calculations were a basis for rejecting the project. But there are obvious benefits. Seeking macro-scale flexibility was prudent.

In contrast, Chapter 2 presents a highway project in New Jersey that was vetoed by the new Governor. The document was adamant about the option, and the governor was not given any flexible alternatives that would have allowed him to find a compromise. Did a bad proposal really go away? Not really. The Department of Transportation and county governments have been making adjustments on other roads, many inefficient, in my opinion, to compensate for what was not built. Almost 40 years after the inflexible Driscoll EIS was completed, I suspect that thoughtful macro-scale planning could have produced a hybrid that the Governor could have approved.

The EIS projects I have worked on have always had a big-picture story, and yet it was rarely captured in the document. This book is full of such examples. The most painful omission was the lack of this bigger picture regarding the destruction of the chemical weapons stockpile on Johnston Island (Chapter 5). By not being candid about the foreign policy need to move chemical weapons ordinance from Germany as soon as possible, the Department of Defense left itself open to charges of environmental injustice (the 1994 presidential order by Bill Clinton did not exist when this EIS was written), and the document itself makes some extremely smart people appear to be disorganized and defensive. Were there other macro-scale options open to the US Army? This was not discussed. A five-page addition to the document would have made the Johnston Island idea appear less insidious to the island nations of the Pacific.

In the case of Sparrows Point (Chapter 4), the lack of a big-picture discussion is embarrassing. There was a great deal of tension around the choice of fossil fuels of any kind at this time. And, as noted in the discussion in Chapter 4, policies passed by state governments were given virtually no credence by the EIS. One need not have a PhD in order to be able to learn within 15 minutes that there were other policy options. Rather than dismissing these, this EIS would have been more credible if it had discussed them seriously. It may be, in fact, that these other options will not succeed, in other words, that they are pie-in-the-sky, but, without acknowledging them, the report leaves itself open to a negative summary judgment on the part of many individuals. As an informational tool, this EIS badly needed a broader perspective about energy. Readers should not have to read a newspaper reporter or blogger to obtain the big picture.

The second level, meso flexibility, is built in to the EIS in the form of the alternative projects or hybrids of them. In Chapter 3, for example, those who wanted to rebuild Ellis Island provided flexible alternatives that could be implemented in stages, depending upon the availability of funds. Chapter 6 describes an exercise in flexibility: a variety of nuclear waste management projects were described, each with different technologies and each based on on-site clean-up objectives, and transfers to and from the site. The final solution is one that focused on cleaning up the most dangerous hazards first, but the document clearly was written in a way to enhance the DOE's capacity to make choices among multiple different options.

Understanding of meso flexibility would be enhanced by a relatively simple matrix that lists each alternative and scores it along the set of impact criteria. I would expect to find some options to be better for water, others for air, and others for socioeconomics. Most of the EISs verbalize these judgments, but it would be good to have them in a single summary table to enhance understanding. This idea is most certainly not novel. But just because it is not new does not mean it should be ignored.

Micro flexibility is in the multiple options that the planning team considers before arriving at the alternatives presented in the EIS. The presentation in Chapter 2 about the light rail system is illustrative. Multiple different routes were considered before the final choice was made, and a variety of public and private decision-makers were involved, in the open and behind closed doors, in those decisions. Without that flexibility to accommodate different stakeholders in different ways, that project would have failed. People I have talked to about the light rail system question some of the decisions, but they are happy it was built, even if they don't agree with every one of the routing decisions. Normal working people cannot be expected to digest dozens of options. Accordingly, the presentation of the EIS should briefly describe the myriad of options available in an appendix for the technically inclined. In fact, this is sometimes the case, and is helpful to those who want to review them.

# What NEPA is not

NEPA is not the mechanism to clean up the water supply, air, or land; control greenhouse gas emissions; or accomplish various other environmental and public health objectives. Its role in water quality, for example, is to require that every federal or federally sponsored project consider the impact of a new project on water quality. However, the federal agency might still choose the alternative that was not the most protective of drinking water, or that did not fully protect water for swimming and fishing.

Even if, miraculously, NEPA requires the results of technical–rational deliberations represented in the EIS to be the basis of agency decisions, NEPA could not lead to the accomplishment of the goal of clean-up of already-contaminated water, air or land, or other such objectives. The United States' environmental policies are almost always incremental. The United States never seems to promulgate a single, comprehensive environmental law that suffices. For example, multiple federal laws were required to address the degradation of water. The Safe Drinking Water Act of 1974 (amended in 1986 and 1996) was born in the wake of fears that the population of New Orleans had higher cancer rates because they were drinking water from the Mississippi River (Greenberg 1983). The Safe Drinking Water Act set standards for drinking water quality and provided for oversight of potable water suppliers (US EPA 2010a).

In order to clean up water bodies and provide potable, fishable, and swimmable waters, Congress passed the Federal Water Pollution Control Act Amendments of 1972, which required classification of all water bodies into water quality categories, and then instituted technologically driven and/or water quality-based approaches to remove contaminants before they were discharged into water bodies (US EPA 2010b).

In the wake of the Love Canal, New York (near Buffalo) hazardous waste disclosures in 1980, Congress passed the Comprehensive Environmental Response Compensation Liability Act (Superfund) (US EPA 2010c), which identified hundreds of sites as threats to public health. The major reason for the majority of these sites to be on the list was water supply contamination (Greenberg and Anderson 1984). The law also set up a mechanism to pay for the management and remediation of the sites.

The fourth law was the Resource Conservation and Recovery Act of 1976 (US EPA 2010e), which gave EPA the authority to manage waste from cradle to grave. Congress also passed the Toxic Substances Control Act of 1976 to manage new potentially toxic substances so that the potential for new pollutants would be reduced (US EPA 2010d).

These laws forced industry, sewage plant operators, and others to directly or indirectly protect water supplies. Water quality is better than it was in 1965, yet all together these laws and billions of dollars invested have not achieved the lofty goals of drinkable, fishable, and swimmable waters. Objectives have been achieved in some places, although in many they have not been. And, like

NEPA, these laws have been criticized for failing to meet their goals, for example, leaving out feedlots from the regulations. My point is that these water-related laws and subsequent regulations directed action at a specific environmental problem. These forced many polluters to take actions that would improve water quality. NEPA was only one part of the effort to improve water quality.

NEPA's ambitious objectives were designed by strategic thinkers who were staring at the mess that was the US environment in the 1960s. As noted in the Preface, as a child living in New York City, it was disheartening to know that nearly all of New York City's sewage was being discharged into surrounding water with no treatment, that an air inversion meant that those of us with respiratory problems would have a difficult time breathing, and that New York City was well on its way to building the largest landfill in the United States. NEPA helped change the mood, but it was unrealistic then, and continues to be unrealistic, to assume that in the United States a single policy document will make a major dent in environmental problems of the early twenty-first century. These problems require a federal government willing to develop a set of policies that together approach a comprehensive agenda.

I am not saying that NEPA should not be significantly adjusted. For example, requiring justification for categorical exclusions and FONSI declarations are excellent ideas. Some bad decisions will be avoided by requiring more information, but only if the agency can be persuaded by more evidence.

## NEPA: accepting shortcomings and building on attributes

I have heard NEPA referred to as a toothless tiger, which it is not because it has teeth if the decision-maker does not muzzle them. I have heard it referred to as a predator feasting on government and business dollars and of little value to decision-makers, which is hard to justify with published data. And I have called it a chameleon capable of blending into agencies' facades. It has been used both as a scientific facade to support decisions that government seemingly was determined to make, and as the rare or only tool available to bring attention to an environmental issue in the absence of Congressional and Executive action. I began writing this chapter in July 2010, and I illustrate NEPA's shortcomings and attributes with two high-profile examples.

### The Gulf Coast oil platform failure and the need for a major addition

The oil platform blowout in the Gulf of Mexico came as a shock to almost everyone, but not to me. I had read a rather obscure risk analysis for the oil and gas leasing program (US Department of the Interior 2007). This sixty-nine-page report presents a brief risk analysis accompanied by dozens of maps. The key data are estimates of blowouts of $\geq 1000$ and $\geq 10,000$ barrels. Estimates of one or more oil spills of $\geq 10,000$ barrels for the central area from platforms were estimated to be 4–6% (US Department of the Interior 2007, Table 1b, p. 52). Pipeline spill estimates were 23–36%. These probabilities were derived by using historical data from 1985–99 and fitting them to a Poisson (rare event) distribution. Other simple models were used to estimate where the spilled oil would go.

I have no reason to doubt this exercise, although I would not have done the analysis in this way. My bigger concern is what happened to the results. When I read 4–6% and 23–36%, I was concerned and assumed that a second-level risk analysis would be conducted. This would have involved detailed engineering analysis of the planned structure and review of human factors to determine how the risk could be reduced far below these estimates.

Reading the FEIS for the Outer Continental Shelf (OCS) program (*ibid.*), I cannot find the equivalent of that analysis, or the specific language that would insist on such steps. The April 2007 document refers to lease stipulations and regulations to minimize loss – language that would reduce the risk. The FEIS talks about adding minimum mitigation measures and monitoring requirements for whales, polar bears, sea turtles, permafrost, and many other conditions. However, the relationship between these activities and risk minimization is not at all clear to me.

I cannot find a scenario analysis in which the worst of the worst scenarios were evaluated, and how these relate to these mitigation and monitoring activities. If the worst-case scenarios had been studied, could detailed engineering alterations and worker behavior changes have prevented the blowout in the Gulf of Mexico? I don't know enough about oil drilling technology to say yes. But I do know that the risk analyses behind the LNG, chemical weapons, and nuclear waste management facilities in this book appear to me to have been far more sophisticated than those on for the offshore oil lease. Would the results of such analyses necessarily have been reflected in the final decisions by agency managers? It may be that, in the end, the decision-maker would have ignored even the most sophisticated risk analysis that could have been prepared.

The Gulf of Mexico blowout is one of many low-probability, high-consequence hazard events that we have suffered during the past decade, including bridge collapses, dam failures, electrical power system failures, oil tanker spills, airplane accidents, terrorist attacks, chemical plant leaks, hurricanes, tornadoes, and on and on. The domains of the unfolding of the event are multilayered, and most importantly involve numerous and pervasive uncertainties. The field of risk analysis was developed to provide information to decision-makers and stakeholders about complex events, and was first systematically applied to nuclear power plants in the United States after the Three Mile Island event.

I understand that we learn more from failures than successes, and perhaps the blowout in the Gulf of Mexico is going to be the last of these events. But it certainly will not be the last serious oil spill, or extremely serious low-probability, high-consequence event. If the US government wants to reduce the likelihood and consequences of these oil-related events, it is going to need to require legislation that mandates not only that average and above-average risks are to be evaluated, but also that a set of plausible worst-case risks need evaluation, including the costs of reducing their likelihood, and the costs of making the system more resilient when an event occurs. This type of analysis leaves an agency with some tough trade-offs, opening it to second-guessing. Does the agency want to add to the cost to dramatically reduce the probability of the worst kinds of events? The data put the pressure directly on the decision-makers, which is where it belongs.

NEPA can accommodate these important major additional risk analyses. However, I am not certain the government agencies are willing to order that they be done; and if they are done, I wonder if the analysis will be buried in an obscure reference two layers below the EIS.

## Global warming and temporary patches

Global warming is perhaps the best current illustration of chronic actions that lead to an accumulation of environmental and public health impacts. Other outcomes of chronic overuse and abuse include chronic water supply and sanitation problems, especially in developing nations, food insecurity, and many other actions that systematically degrade the quality of life of people in general, especially impoverished people. The weight of scientific evidence shows that $CO_2$ emissions contribute to global warming. The most effective mechanism to address the issue is a single or multiple climate change law(s). But there is none in the United States. Meanwhile, there are always Presidential Executive Orders and NEPA to at least send a message, flush out the inevitable opposition, and initiate some action.

In January 2010, Council on Environmental Quality (CEQ) chairwoman Nancy Sutley (CEQ 2010; see also Straub 2010a,b) noted that NEPA could be used to require new federal government initiatives to include greenhouse gas evaluations. On February 18, 2010, CEQ issued draft guidance on climate change and greenhouse gas emissions. The guidance calls for estimates of greenhouse gas emissions, consideration of mitigative measures and conservation, and several other steps. It gives some examples about considering the location of infrastructure in places that could be flooded. The draft guidance ends with a set of questions for public response. I would characterize it as vague.

The responses were as anticipated from the American Association of Home Builders (L. Mark, National Association of Home Builders, "Draft NEPA guidance on consideration of the effects of climate change and greenhouse gas emissions", letter to Nancy Sutley, May 21, 2010) and from Congressional opponents (Straub 2010b) that the guidance is too vague and too costly to implement during a recession. More letters will follow and the guidance will be adjusted. For me, this guidance brings the life history of NEPA back to its origins when there was no clean water or air legislation, or the other two dozen key laws that have markedly improved the US environment. In early 1970, NEPA sent the message and required some action. Forty years later, it is the scout for other, more powerful legislation that may or not be passed in the near future.

Is applying risk analysis systematically across agencies that are responsible for low-probability, high-consequence events, and requiring additional thinking and analysis documents about global climate change and other chronic impacts, adding to an endless cycle of potentially significant but costly analysis that will be ignored? Readers who believe that NEPA already has gone too far to embrace the technical–rational planning decision-making process will find limited solace in these ideas. Opening up more freewheeling studies has merit only if they are to be used seriously by decision-makers before decisions are made, and/or by evaluators to assess their value in the performance of decision-makers. In the end, I think we want wise decisions and reasonable decision-making processes. NEPA's ideas remain an inspiration, and EIS planning documents have a mixed record. Yet, for me, they have become even more important during a period of economic uncertainty unprecedented for over half a century, when the pressure to produce immediate benefits and worry about impacts later is prevalent. Reiterating, I believe that the three major attributions required by EISs are important in, of, and by themselves, and they can sometimes partly compensate for the absence of other major environmental laws. These attributes would be enhanced by the suggestions offered in this chapter. In the end, however, open-minded decision-makers and staff are necessary before the current requirements will be put to use on behalf of society.

# Bibliography

••••••••••••••••••••••••••••••••••••••••••••••••••••••••••••

Alfano, P. Undated. *NEPA at 40: Procedure or Substance.* www.eli.org/pdf/seminars/ NEPA/Alfano.NEPA.pdf

Allport GW. 1954. *The Nature of Prejudice.* Boston, MA: Addison-Wesley.

Amin A. 2002. *Ethnicity and the Multicultural City: Living with Diversity.* London: Department for Transport, Local Government and the Regions and the ESRC Cities Initiative.

Andelman D. 1971. New Montauk route may be going nowhere. *The New York Times* November 14, p. A1.

Andrews RL. 1976. *Environmental Policy and Administrative Change: Implementation of the National Environmental Policy Act.* Lexington, MA: Lexington Books.

Atkinson S, Canter L, Ravan M. 2006. The influence of incomplete or unavailable information on environmental impact assessment in the USA. *Environmental Impact Assessment Review* 26(5): 448–467.

Baecher G, Gross J, McCusker K. 1975. *Balancing Apples and Oranges: Methodologies for Facility Siting Decisions.* Laxenberg, Austria: International Institute for Applied Systems Analysis.

Baehr G. 1989a. Conrail, NJ Transit sign deal for Hudson trolley. *The Star-Ledger* June 9.

Baehr G. 1989b. Waterfront transit panel formed to study improvements along the Hudson. *The Star-Ledger* August 22.

Baehr G. 1990. Group criticizes plan for Hudson 'busway,' *The Star-Ledger* October 18.

Baehr G. 1991. NJ Transit lays out plan to lay down rail for Hudson waterfront trolley. *The Star-Ledger* October 23.

Baehr G. 1993. NJ Transit approves trolley line to serve the Hudson waterfront. *The Star-Ledger* February 11.

Baehr G. 1995. Waterfront trolley on track with final approvals. *The Star-Ledger* March 29.

Baker P, Bilefsky D. 2010. Russia and U.S. sign nuclear arms reduction pact. *The New York Times* April 9, p. A8.

Bardach E. 1977. *The Implementation Game: What Happens After a Bill Becomes a Law.* Cambridge, MA: MIT Press.

Berlin R, Stanton C. 1989. *Radioactive Waste Management.* New York: Wiley.

Best J. 1972. *The National Environmental Policy Act as a Full Disclosure Law.* Cornell University Energy Project. Washington DC, National Technical Information Service.

Blair W. 1971a. Controversy grows over Teton River dam project. *The New York Times* August 8, p. 48.

Blair W. 1971b. Court wholesale of offshore leases. *The New York Times* December 17 p. 6.

Borough of Tenafly 2009. *Newsletter* 50, February. Tenafly, NJ: Borough of Tenafly.

Bregman J. 1999. *Environmental Impact Statements*. 2nd edn. Boca Raton, FL: Lewis.

Brookings Institution. 1998. *50 Facts about U.S. Nuclear Weapons*. Washington, DC: The Brookings Institution. www.brookings.edu/projects/archive/nucweapons/ 50.aspx

Bryant N. 1971. Conservations hail court decision halting work on Arkansas dam. *The New York Times* March 14, p. S25.

Buck B, Hobbs M, Kaiser A, Lang S, Montero D, Romines K, Scott Y. 2003. Immigration and immigrants: trends in American public opinion, 1964–1999. *Journal of Ethnic and Cultural Diversity in Social Work* 12: 73–90.

Burchell R, Dolphin W, Galley C. 2000. *The Costs and Benefits of Alternative Growth Patterns: The Impact Assessment of the New Jersey State Plan*. New Brunswick, NJ: EJ Bloustein School of Planning and Public Policy.

Burchell R, Lowenstein G, Dolphin W, Galley C, Downs A, Seskin S, Gray Still K, Moore T. 2002. *Costs of Sprawl – 2000*. Transportation Research Board, National Research Council. Washington DC: National Academy Press.

Burger J. 2000. Contaminated Department of Energy facilities and ecosystems: weighing the ecological risks. *Journal of Toxicology and Environmental Health* A63: 383–95.

Burger J, Leschine T, Greenberg M, Karr J, Gochfeld M, Powers C. 2003. Shifting priorities at the Department of Energy's bomb factories: protecting human and ecological health. *Environmental Management* 31(2): 57–167.

Burns P, Gimpel J. 2000. Economic insecurity, prejudicial stereotypes, and public opinion on immigration policy. *Political Science Quarterly* 115: 201–225.

Caldwell L. 1982. *Science and the National Environmental Policy Act*. Tuscaloosa, AL: University of Alabama Press.

Caldwell L. 1989. NEPA revisited: a call for a constitutional amendment. *Environmental Focus* November/December: 18–22.

Caldwell L. 1998. *The National Environmental Policy Act*. Harvard Environmental. Bloomington, IN: Indiana University Press.

California Energy Commission. Undated. *Liquefied Natural Gas Safety*. www.energy.ca.gov/lng/safety.html

Canter L. 1977. *Environmental Impact Assessment*. New York: McGraw-Hill.

Carson R. 1962. *Silent Spring*. Boston, MA: Houghton Mifflin.

Carter L. 1987. *Nuclear Imperatives And Public Trust: Dealing With Radioactive Waste*. Baltimore, MD: Johns Hopkins University Press.

CEQ. 1997a. *The National Environmental Policy Act, A Study of its Effectiveness after Twenty-Five Years*. Washington, DC: Council on Environmental Quality, Executive Office of the President. http://ceq.hss.doe.gov/nepa/nepa25fn.pdf

CEQ. 1997b. *Environmental Justice Guidance Under the National Environmental Policy Act*. Washington, DC: Council on Environmental Quality, Executive Office of the President. http://ceq.hss.doe.gov/nepa/regs/ej/justice.pdf

CEQ. 2007. *A Citizen's Guide to the NEPA*. Washington, DC: Council on Environmental Quality, Executive Office of the President. ceq.hss.doe.gov/nepa/ Citizens_Guide_Dec07.pdf

CEQ. 2010. *Steps to Modernize and Reinvigorate NEPA*. Washington, DC: Council on Environmental Quality, Executive Office of the President. www.whitehouse. gov/administration/eop/ceq/initiatives/nepa

Chaker A, El-Fadik K, Chamas L, Hatjan B. 2006. A review of strategic environmental assessment in 12 selected countries. *Environmental Impact Assessment Review* 26: 15–56.

Chandler C, Yung-mei T. 2001. Social factors influencing immigrant attitudes: an analysis of data from the General Social Survey. *Social Science Journal* 38: 177–188.

CMA. 2010. *Creating a Safer Tomorrow*. Edgewood, MD: Chemical Materials Agency.

Cheremisinoff P, Morresi A. 1977. *Environmental Assessment and Impact Statement Handbook*. Ann Arbor, MI: Ann Arbor Science Publishers.

Cheslow J. 1972. Manalapan battles road plan. *The New York Times* November 26, p. 143.

Citrin J, Green D, Muste C, Wong C. 1997. Public opinion toward immigration reform: the role of economic motivations. *Journal of Politics* 59: 858–881.

Committee on Foreign Relations. 1996. *Executive Report 104-33*. Washington, DC: US Senate.

Committee on Resources. 2006. *NEPA: Lessons Learned and Next Steps. Oversight Hearing*. US House of Representatives, 109th Congress, First Session. November 17, 2005. Serial No. 109-37. Washington, DC: US Government Printing Office. http://ftp.resource.org/gpo.gov/hearings/109h/24682.txt

Cullingsworth JB. 1993. *The Political Culture of Planning. American Land Use in Comparative Perspective*. New York: Routledge.

Cumbler J. 2005. *Northeast and Midwest United States: An Environmental History*. Santa Barbara, CA: ABC-CLIO.

Dawson G. 1973. Turnpike authority seeks to answer objections to new spur; cost estimates bonds snapped up. *The New York Times* June 10, p. 83.

Decker P. 2004. *The Utes Must Go*. Golden, CO: Fulcrum Publishing Company.

Dinar A, Seidel P, Olem H, Jorden V, Duda A, Johnson R. 1995. *Restoring and Protecting the World's Lakes and Reservoirs*. World Bank Technical Paper 289. Washington, DC: World Bank, pp. 6–7.

Draper E. 2006. Animas-La Plata Project nearing halfway point. *Denver Post* July 20. www.denverpost.com/ci_4072305

Dreyfus D, Ingram H. 1976. The National Environmental Policy Act: a view of the intent and practice. *Natural Resources Journal* 26(2): 243–262.

Dweck J, Wochner D, Brooks M. 2006. Liquefied natural gas (LNG) litigation after the Energy Policy Act of 2005: State powers in LNG terminal siting. *Energy Law Journal* 27(45): 482–485.

Ehrlich P. 1968. *The Population Bomb*. New York: Ballantine.

EIA. 2009. *Annual Energy Outlook 2009, With Projections to 2030*. Washington, DC: Energy Information Administration, National Energy Information Center. www.eia.gov/oiaf/aeo/trends.html

Ellison B. 1999. Environmental management and the new politics of Western water. The Animas-La Plata project and implementation of the Endangered Species Act. *Environmental Management* 23(4): 429–439.

Espenshade T. 1997. New Jersey in comparative perspective. pp. 1–31 in *Keys to Successful Immigration*, edited by TJ Espensade. Washington, DC: Urban Institute Press.

Espenshade T, Hempstead K. 1996. Contemporary American attitudes toward U.S. immigration. *International Migration Review* 30: 535–570.

Fairfax S. 1978. A disaster in the environmental movement. *Science* 199: 743–748.

FERC. 2008a. *Draft Environmental Impact Statement: Sparrows Point LNG Terminal and Pipeline Project.* Washington, DC: Federal Energy Regulatory Commission.

FERC. 2008b. *Final Environmental Impact Statement on Sparrows Point LNG and Mid-Atlantic Express Pipeline Project.* Washington, DC: Federal Energy Regulatory Commission.

FERC. 2009. *FERC approves AES Sparrows Point LNG terminal, Mid-Atlantic Express Pipeline.* January 15, www.ferc.gov/media/news-releases/2009/2009-1/01-15-09-C-1.pdf

Fischer T. 2007. *Theory and Practice of Strategic Environmental Assessment – Toward a More Systematic Approach.* London: Earthscan.

Fishman R. 2000. American metropolis at century's end: past and future influences. *Housing Policy Debate* 11(1): 199–213.

Flippen J. 2000. *Nixon and the Environment.* Albuquerque, NM: University of New Mexico Press.

Forbes HD. 1997. *Ethnic Conflict: Commerce, Culture and the Contact Hypothesis.* New Haven, CT: Yale University Press.

Franklin B. 1971. Senator scores U.S. coal leases. *The New York Times* November 7, p. 83.

Galisteo Consulting Group. 2002. *Social, Cultural, Economic Impact Assessments: A Literature Review.* Washington, DC: Office of Emergency and Remedial Response, USEPA.

GAO. 1996. *Chemical Weapons Stockpile, Emergency Preparedness in Alabama is Hampered by Management Weakness*, Publication 96-150. Washington, DC: US General Accounting Office /NSAID.

GAO. 1997. *Chemical Weapons Stockpile: Changes Needed in the Management of the Emergency Preparedness Program*, Publication 97-91. Washington, DC: US General Accounting Office /NSAID.

GAO. 2001. *FEMA and Army Must be Proactive in Preparing States for Emergencies*, Publication 01-850. Washington, DC: US General Accounting Office.

GAO. 2007. *Maritime Security: Federal Efforts Needed to Address Challenges in Preventing and Responding to Terrorist Attacks on Energy Commodity Tankers.* GAO-08-141, December 10. Washington, DC: US Government Accountability Office.

General Assembly of the UN. 1992. *Convention on the Prohibition of the Development, Production, Stockpiling and Use of Chemical Weapons and on Their Destruction.* New York: United Nations.

Gilbertson M. 2009. Energy Parks Initiative. Presentation, August 18 at Savannah River Site. www.ncsl.org/documents/environ/MGilbertson0609.pdf

Glazer N, Moynihan D. 1970. *Beyond the Melting Pot: The Negroes, Puerto Ricans, Jews, Italians, and Irish of New York City.* Cambridge, MA: MIT Press.

Goodchild M. 2007. Citizens as sensors: the world of volunteered geography. *GeoJournal* 69: 211–221.

Greenberg M. 1983. *Urbanization and Cancer Mortality.* New York: Oxford University Press.

Greenberg M. 2003. Public health, law, and local control: Destruction of the US chemical weapons stockpile. *American Journal of Public Health* 93(8): 1222–1226.

Greenberg M. 2008. *Environmental Policy Analysis and Practice*. New Brunswick, NJ: Rutgers University Press.

Greenberg M. 2009. NIMBY, CLAMP and the location of new nuclear-related facilities: U.S. national and eleven site-specific surveys. *Risk Analysis* 29(9): 1242–1254.

Greenberg M, Anderson R. 1984. *Hazardous Waste Sites: The Credibility Gap*. New Brunswick, NJ: Center for Urban Policy Research.

Greenberg M, Cidon M. 1997. Broadening the definition of environmental equity: a framework for states and local governments. *Population Research and Policy Review* 16: 397–413.

Greenberg M, Belnay G, Cesanek W, Neuman N, Shepherd G. 1978. *A Primer on Industrial Environmental Impact*. New Brunswick, NJ: Rutgers University, Center for Urban Policy Research.

Greenberg M, Lowie K, Krueckeberg D, Mayer H, Simon D. 1997. Bombs and butterflies: a case study of the challenges of post cold-war environmental planning and management for the United States nuclear weapons sites. *Journal of Environmental Planning and Management* 40: 739–750.

Greenberg M, Miller KT, Frisch M, Lewis D. 2003. Facing an uncertain economic future: environmental management spending and rural regions surrounding the U.S. DOE's nuclear weapons facilities. *Defence and Peace Economics* 14(1): 85–97.

Greenberg M, Lahr M, Mantell N. 2007. Understanding the economic costs and benefits of catastrophes and their aftermath: a review and suggestions for the US federal government. *Risk Analysis* 27(1): 83–96.

Greenberg M, Lowrie K, Hollander J, Burger J, Powers C, Gochfeld, M. 2008. Citizen board issues and local newspaper coverage of risk, remediation, and environmental management: six U.S. nuclear weapons facilities. *Remediation* Summer 79–90.

Greenberg M, West B, Lowrie K, Mayer H. 2009. *The Reporter's Handbook on Nuclear Materials, Energy, and Waste Management*. Nashville, TN: Vanderbilt University Press.

Greenspan A. 2003 Natural gas supply and demand issues. Testimony before the House Energy and Commerce Committee. In Parfomak P, Vann A. 2008. *CRS Report for Congress. Liquefied Natural Gas (LNG) Import Terminals: Siting, Safety, and Regulation*.

Hagman, D. 1974. NEPA's progeny inhabit the states – were the genes defective? *Urban Law Annual* 7: 3–56.

Haubert J, Fussell E. 2006. Explaining pro-immigrant sentiment in the U.S.: social class, cosmopolitanism, and perceptions of immigrants. *International Migration Review* 40: 489–507.

Hennelly B. 2009. Salazar tours Ellis Island and Statue of Liberty. *WNYC News* January 24. www.wnyc.org/news/articles/121996

Hewstone M, Rubin M, Willis H. 2002. Intergroup bias. *Annual Review of Psychology* 53: 575–604.

Hill G. 1970. Pollution: It's time to "get cracking" on reforms. *The New York Times* August 16, p. 133.

Hirschman C. 2005. Immigration and the American century. *Demography* 42: 595–620.

Houck O. 2000. Is that all? A review of *The National Environmental Policy Act, An Agenda for the Future*, by Lynton Keith Caldwell. *Duke Environmental Law and Policy Forum* 11: 178–180.

Hurley A. 1995. *Environmental Inequalities: Class, Race, and Industrial Pollution in Gary, Indiana, 1945–1980.* Chapel Hill, NC: University of North Carolina Press.

Hurst C. 2008. *The terrorist threat to liquefied natural gas: fact or fiction?* Washington, DC: Institute for the Analysis of Global Security.

Jain R, Urban L, Stacey G, Balbach H. 2002. *Environmental Impact Analysis.* 2nd edn. New York: McGraw Hill.

Janson D. 1974. 18,000 rally in Trenton for more jobs. *The New York Times* July 30, p. 69.

*Jersey Journal.* 1997. Plan to make good use of western light rail route. Editorial, April 23.

Kaplan S, Garrick BJ. (1981). On the quantitative definition of risk. *Risk Analysis* 1(1): 11–27.

Kaufman H. 1997. The role of NEPA in sustainable development. pp. 313–320 in Clark R, Canter L (eds) *Environmental Policy and NEPA: Past, Present and Future.* Boca Raton, FL: St Lucie Press.

Kenworthy, E. 1970a. Hart prods Nixon on Environment Act. *The New York Times* November 19, p. 16.

Kenworthy, E. 1970b. Administration assailed by Muskie on the SST. *The New York Times* December 2, p. 74.

Kenworthy, E. 1970c. SST opponents charge administration suppressed criticism. *The New York Times* December 11, p. 37.

Kenworthy E. 1971a. It was quite a week for Ruckelshaus watchers. *The New York Times* March 21, p. E3.

Kenworthy E. 1971b. Resort plan sparks Montana controversy. *The New York Times* May 31, p. 6.

Kenworthy E. 1971c. F.P.C. hears foes of power plant. *The New York Times* November 12, p. 18.

Key VO Jr. 1949. *Southern Politics in State and Nation.* New York: Alfred A. Knopf.

Keysar E. 2005. Procedural integration and support of environmental policy objectives: implementing sustainability. *Journal of Environmental Planning and Management* 48(4): 549–569.

Keysar E, Steinemann A. 2002. Integrating environmental impact assessment with master planning: lessons from the U.S. Army. *Environmental Impact Assessment Review* 22: 583–609.

Kreske D. 1996 *Environmental Impact Statements: A Practical Guide for Agencies, Citizens, and Consultants.* New York, NY: Wiley.

Kubiszewski I. 2006. Nuclear Waste Policy Act of 1982. In Cleveland CL (ed.), *Encyclopedia of Earth.* Washington, DC: Environmental Information Coalition, National Council for Science and the Environment.

Lambright W, Gereben A, Cerveny L. 1998. The Army and chemical weapons destruction: implementation in a changing context. *Policy Studies Journal* 26: 703–718.

Lapham S. 1993. *We the American Foreign Born.* Washington, DC: US Department of Commerce, US Census Bureau.

Lapinski J, Peltola P, Shaw G, Yang A. 1997. The polls – trends: immigrants and immigration. *Public Opinion Quarterly* 61: 356–383.

Lapolla M, Suszka T. 2005. *Images of America: The New Jersey Turnpike.* Chicago, IL: Arcadia.

Lapp R. 1971. Nuclear tests: Alaskans again ask uneasy questions. *The New York Times* April 18, p. E6.

Lawrence D. 2007a. Impact significant determination – defining an approach. *Environmental Impact Assessment Review* 27: 730–754.

Lawrence D. 2007b. Impact significant determination – back to basics. *Environmental Impact Assessment Review* 27: 755–769.

Lawrence D. 2007c. Impact significant determination – pushing the boundaries. *Environmental Impact Assessment Review* 27: 770–778.

Lindstrom M, Smith D. 2001. *National Environmental Policy Act: Judicial Misconstruction, Legislated Indifference, and Executive Neglect.* College Station, TX: Texas A&M University Press.

Ling K. 2010. Court ruling imperils Baltimore LNG proposal. *The New York Times* January 4. www.nytimes.com/gwire/2010/01/04/04greenwire-court-ruling-imperils-baltimore-lng-proposal-76939.html

Liroff R. 1976. *A National Policy for the Environment – NEPA and its Aftermath.* Bloomington, IN: Indiana University Press.

LNGpedia. 2009. EIA sharply raises U.S. LNG import estimate. *LNGpedia* April 16. www.lngpedia.com

MacFarlane A, Ewing R (eds). 2006. *Uncertainty Underground: Yucca Mountain and the Nation's High Level Nuclear Waste.* Cambridge, MA: MIT Press.

MacFarquhar N. 1995. With wide gulf separating factions, battle over Ellis Island bridge resumes. *The New York Times*, May 8. www.nytimes.com/keyword/ellis-island/4

Maddox J. 1972. *The Doomsday Syndrome.* New York: McGraw-Hill.

Malone N, Baluja K, Costanzo J, Davis C. 2003. *The Foreign-Born Population: 2000.* Washington, DC: US Census Bureau.

McCool D. 1994. Utah and the Ute tribe are at war. *High Country News* June 27. www.hcn.org/issues/9/285/print_view

McMenamin J. 2006. Sparrows Point LNG plant raises fears: potential for explosives, terrors threats, restricted access concerns residents. *The Baltimore Sun* June 6. http://articles.baltimoresun.com/2006-06-06/news/0606060060_1_sparrows-point-baltimore-county-lng

McMillan P. 2005. *The Ruin of J. Robert Oppenheimer.* New York: Viking.

McShane C. 1994. *Down the Asphalt Path: The Automobile and the American City.* New York: Columbia University Press.

Melosi M. 2001. *Effluent America: Cities, Industry, Energy, and the Environment.* Pittsburgh, PA: University of Pittsburgh Press.

Miniwatts Marketing Group. 2010. *Internet Usage Statistics for the Americas.* www.internetworldstats.com/stats2.htm

Montoya S. 2006. *Oral History.* Las Vegas, NV: Colorado River Water Users Association. www.crwua.org/AboutUs/OralHistory.aspx

Munro NB, Talmage SS, Griffin GD, Waters LC, Watson AP, King JF, Hauschild V. 1999. The sources, fate, and toxicity of chemical warfare agent degradation products. *Environmental Health Perspectives* 107: 933–974.

National Park Service. 2005. *Ellis Island National Monument, Final General Management Plan and Environmental Impact Statement.* Washington, DC: National Technical Information Service.

National Park Service. 2008. *Governors Island National Monument, Final General Management Plan and Environmental Impact Statement.* Washington, DC: National Technical Information Service.

National Research Council. 1998. *Using Supercritical Water Oxidation to Treat Hydrolysate from VX Neutralization.* Committee on Review and Evaluation of the Chemical Stockpile Disposal Program. Washington, DC: National Academy Press.

National Research Council. 2000a. *Integrated Design of Alternative Technologies for Bulk-Only Chemical Agent Disposal Facilities.* Committee on Review and Evaluation of the Chemical Stockpile Disposal Program. Washington, DC: National Academy Press.

National Research Council. 2000b. *A Review of the Army's Public Affairs Efforts in Support of the Chemical Stockpile Disposal Program.* Committee on Review and Evaluation of the Chemical Stockpile Disposal Program. Washington, DC: National Academy Press.

National Research Council. 2001. *Occupational Health and Monitoring at Chemical Agent Disposal Facilities.* Committee on Review and Evaluation of the Chemical Stockpile Disposal Program. Washington, DC: National Academy Press.

National Safety Council. 2001. *A Reporter's Guide to Yucca Mountain.* Washington, DC: Environmental Health Center, National Safety Council. http://downloads. nsc.org/PDF/yuccapdf.pdf

New Jersey Turnpike Authority. 1972. *Governor Alfred E. Driscoll Expressway. Environmental Impact Statement.* Trenton, NJ: New Jersey Turnpike Authority.

*New York Daily News.* 2009. Daily News contest winner letters to the Statue of Liberty. July 5. www.nydailynews.com/ny_local/2009/07/05/2009-07-05_daily_ news_contest_winners_let.html

*New York Times.* 1970. U.S. said to delay ecology reports. November 14, p. 22

*New York Times.* 1971a. Oil spill feared in drilling plans. July 24, p. 50.

*New York Times.* 1971b. Morton studying Alaskan pipeline. October 5, p. 18.

*New York Times.* 1971c. Suits seek to bar A-power station. November 7, p. 81.

*New York Times.* 1972. Reactions mixed on Turnpike plan. December 16, p. 66.

*New York Times.* 1973a. Turnpike official pledges study of revised Driscoll expressway. September 13, p. 98.

*New York Times.* 1973b. New Jersey briefs; Turnpike lets Driscoll contract. November 22, p. 78.

NJ Transit. 1992. *Draft Environmental Impact Statement.* November S-7. Newark, NJ: New Jersey Transit.

NJ Transit. 2007. *NJ Transit announces record ridership, talks Secaucus parking.* Newark, NJ: New Jersey Transit. http://blog.tstc.org/2007/12/13/nj-transit-announces-record-ridership-talks-secaucus-parking/

Office of Environmental Management. 1995. *Closing the Circle on Splitting the Atom.* Washington, DC: US Department of Energy.

Office of Environmental Management. 1997. *Accelerated Cleanup: Focus on 2006.* Washington, DC: US Department of Energy.

Office of Environmental Management. 2010a. *About the EM Site-specific Advisory Board.* Washington, DC: US Department of Energy.

Office of Environmental Management. 2010b. *Savannah River Site – Citizens Advisory Board.* Washington, DC: US Department of Energy. www.srs.gov/general/ outreach/srs-cab/srs-cab.html

Oregon State University. 1973. *How Effective are Environmental Impact Statements?* PB-230-702. Springfield, VA: National Technical Information System.

Ortolano L, ed. 1973. *Analyzing the Environmental Impacts of Water Projects.* Alexandria, VA: U.S. Army Corps of Engineers, Institute for Water Resources.

Padilla Y. 1997. Immigrant policy: issues for social work practice. *Social Work* 42: 595–606.

Pantoja A. 2006. Against the tide? Core American values and attitudes towards US immigration policy in the mid-1990s. *Journal of Ethnic and Migration Studies* 32: 515–531.

Parfomak P, Vann A. 2008. *Liquefied Natural Gas (LNG) Import Terminals: Siting, Safety, and Regulation.* CRS Report for Congress. Washington, DC: Congressional Research Service.

Partidario M. 2000. Elements of an SEA framework – improving the added value of SEA. *Environmental Impact Assessment Review* 20: 647–663.

Perlstein R. 2008. *Nixonland.* New York: Scribner.

Person J. 2006. Theoretical reflections on the connection between environmental assessment methods in conflict. *Environmental Impact Assessment Review* 26: 605–613.

Pettigrew T, Tropp L. 2005. A meta-analytic test of intergroup contact theory. *Journal of Personality and Social Psychology* 90: 751–783.

Phalon R. 1975. Investors concerned over Pike's toll's fare. *The New York Times* October 16.

Popper D, Popper F. 1987. The Great Plains: from dust to dust. *Planning* 53: 12–18.

Popper F. 1992. Thinking globally, acting regionally. *Technology Review* 95(3): 47–53.

Popper F. 1993. Rethinking regional planning. *Society* 30(6): 46–54.

Pressman J, Wildavsky A. 1973. *How Great Expectations in Washington are Dashed in Oakland, or why it's amazing that federal programs work at all, this being the saga of the Economic Development Administration as told by two sympathetic observers who seek to build morals on a foundation of ruined hopes.* Berkeley, CA: University of California Press.

Rabaska. 2008. *The Future the Natural Way.* Lévis Quebec: Rabaska Project. www.rabaska.net/safety

Reid H. 2007. Yucca Mountain (and related press releases). Website of Harry Reid, US Senator for Nebraska. reid.senate.gov/issues/yucca.cfm

Roberts G, ed. 1999. *American Cities and Technology. Wilderness to Wired City.* New York: Routledge.

Robinson JM. 2005. Testimony of J. Mark Robinson, Director, Office of Energy Project, Federal Energy Regulatory Commission before the Committee on Environment and Public Works, United States Senate, May 25, 7–8.

Rockwell W. 2006 [1956]. *The Utes: A Forgotten People.* Montrose, CO: Western Reflections.

Rodenbaugh D. 2008. A-LP project 97% complete. *The Durango Herald* October 17. www.waterinfo.org/node/2454.

Rodenbaugh D. 2009. Fill 'er up. *The Durango Herald.* May 3. www.waterinfo.org/node/4870

Rosen A. 2007. Sparrows Point LNG terminal decision due within the week. *The Daily Record* January 11.

Rotunda R. 1998. The chemical weapons convention: political and constitutional issues. *Constitutional Commentary* (online subscription) 15: 131. http://web.Lexis-nexis.com/universe

Ruddy T, Hilty L. 2008. Impact assessment and policy learning in the European Commission. *Environmental Impact Assessment Review* 28: 90–105.

Ruppersberger D. 2009. Maryland delegation calls on FERC to delay LNG decision until Obama Administration takes office. http://dutch.house.gov/list/press/md02_ruppersberger/09_01_14_MD_letter_to_FERC.html

Save Ellis Island. 2006. *Ellis Island Institute: An Immigration Icon Reborn for the Twenty-First Century, Mission, Goals and Vision.* Budd Lake, NJ: Ellis Island Institute.

Save Ellis Island. 2007. *Ellis Island Institute: An Immigration Icon Reborn for the Twenty-First Century, Programs and Operation Strategy.* Budd Lake, NJ: Ellis Island Institute.

Save Ellis Island. 2009. After tour with Interior Secretary, Senator Menendez expresses optimism on reopening of Lady Liberty's Crown, restoration of Ellis Island. Budd Lake, NJ: Ellis Island Institute.

Scheve K, Slaughter M. 2001. Labor market competition and individual preferences over immigration policy. *Review of Economics and Statistics* 83: 133–145.

Schneider S. 2007. Anti-immigrant attitudes in Europe: outgroup size and perceived ethnic threat. *European Sociological Review* 24: 53–67.

Schultz S. 2007. Bill proposed to boost states power to block LNG plants. *Baltimore Business Journal* April 24. http://baltimore.bizjournals.com/baltimore/stories/2007/04/23/daily15.html

Schwartz S. 1998. *Atomic Audit: The Costs and Consequences of U.S. Nuclear Weapons since 1940.* Washington, DC: The Brookings Institution.

Sexton K, Linder S. 2010. The role of cumulative risk assessment in decisions about environmental justice. *International Journal of Public Health* 7: 4037–4049.

Shaw T. 1998. Supreme court decides ownership of historic Ellis Island. Oxford, MS: University of Mississippi, Mississippi–Alabama Sea Grant Legal Program. www.olemiss.edu/orgs/SGLC/MS-AL/18.4/ellis.htm

Simmons D. 2000. *The Ute Indians of Utah, Colorado, and New Mexico.* Boulder, CO: University Press of Colorado.

Snell T, Cowell R. 2006 Scoping in environmental impact assessment: balancing precaution and efficiency. *Environmental Impact Assessment Review* 26: 359–376.

State of Maryland. 2006. Special Joint Meeting of the Sport Fishery Advisory Commission and Tidal Fishers Advisory Commission, State of Maryland. February 22. Annapolis, MD: Maryland Department of Natural Resources.

State of New Jersey. 1972. *New Jersey Administrative Code.* Guidelines for Environmental Impact Statement, New Jersey Turnpike Extension, Title Seven, Chapter 1A. Trenton, NJ: State of New Jersey.

Steinemann, A. 2001 Improving alternatives for environmental impact assessment. *Environmental Impact Assessment Review* 21: 3–21.

Stradling D. 1999. *Smokestacks and Progressives: Environmentalists, Engineers, and Air Quality in America, 1881–1951.* Baltimore, MD: Johns Hopkins University Press.

Straub N. 2010a. 'No basis' for excluding climate impacts from NEPA reviews, CEQ says. *The New York Times*, January 15. www.nytimes.com/gwire/2010/01/15/15greenwire-no-basis-for-excluding-climate-impacts-from-ne-77722.html

Straub N. 2010b. Senate Republicans move to bar NEPA analyses of climate change impacts. *The New York Times*, April 10. www.nytimes.com/gwire/2010/04/20/20greenwire-senate-republicans-move-to-bar-nepa-analysis-o-53404.html

Stryker S. 1987. The vitalization of symbolic interactionism. *Social Psychology Quarterly* 50: 83–94.

Sullivan J. 1979. N.J. transportation bonds look good, for a change; a federal–state match, *The New York Times* October 21, p. E6.

Sullivan R. 1973. Byrne's opposition blocks Jersey Turnpike extension; objection form Byrne blocks extension of Turnpike in Jersey, $50-million loss seen. *The New York Times* December 18, p. 1.

Sullivan R. 1974. Spur for Turnpike is blocked by court. *The New York Times* June 18, p. 81.

Sullivan R. 1975. Driscoll funeral held in Haddonfield, but Byrne is absent at request of family. *The New York Times* March 13, p. 83.

Sutley N. 2010. *Draft NEPA Guidance on Consideration of the Effects of Climate Change and Greenhouse Gas Emissions*. Washington, DC: Council on Environmental Quality, Executive Office of the President, February 18.

Tang D, Bright E, Brody S. 2009. Evaluating California local land-use plan's environmental impact reports. *Environmental Impact Assessment Review* 29: 6–106.

Tarr J, ed. 2003. *Devastation and Renewal: An Environmental History of Pittsburgh and its Region*. Pittsburgh, PA: University of Pittsburgh Press.

Taylor JS. 1984. *Making Bureaucracy Think: The Environmental Impact Statement Strategy of Administrative Reform*. Stanford, CA: Stanford University Press.

Therival R. 2004. *Strategic Environmental Assessment in Action*. London: Earthscan

Time Magazine. 1990. The browning of America. April 9, 135(15).

Torres A. 1997a. Chainsaw massacre: hundreds of bayonne trees killed for rail line. *The Jersey Journal* February 1.

Torres A. 1997b. Rat invasion: light rail construction moves rodents into back yards. *The Jersey Journal* June 27.

Torres A. 1998a. Buried treasure? Light rail prep work in area rich in history. *The Jersey Journal* June 8.

Torres A. 1998b. Light rail racket: construction noise plagues Paulus Hook. *The Jersey Journal* October 21.

Torres A. 1999. Health department shuts down light rail site. *The Jersey Journal* January 30.

Train R. 1971. Toward a better environment. *The New York Times* February 11, p. 45.

Tzoumis K. 2007. Comparing the quality of draft environmental impact statements by agencies in the United States since 1998 to 2004. *Environmental Impact Assessment Review* 27, 26–40.

Tzoumis K, Feingold L. 2000. Looking at the quality of draft impact statements in the United States overtime: have ratings improved? *Environmental Impact Assessment Review* 20, 1–22.

Ullmann D. 2009. Opponents ready for overtime in Sparrows Point LNG plant dispute. January 15. www.highbeam.com/doc/1P2-19752125.html

US Army. 1983. *Final Environmental Impact Statement, Johnston Atoll Chemical Agent Disposal System (JACADS)*. Fort Shafter, Hawaii: Toxic and Hazardous Materials Agency.

US Army. 1988. *Chemical Stockpile Disposal Program Full Programmatic Environmental Impact Statement*, Vols 1–3. Aberdeen Proving Ground, MD: US Army.

US Army. 1990a. *Johnston Atoll Chemical Agent Disposal System, Final Second Supplemental Environmental Impact Statement for the Storage and Ultimate Disposal of the European Chemical Munition Stockpile*, Vol. 1. Aberdeen Proving Ground, MD: US Army.

US Army. 1990b. *Comments on Johnston Atoll Chemical Agent Disposal System, Final Second Supplemental Environmental Impact Statement*, Vol. 2. Aberdeen Proving Ground, MD: US Army.

US Army. 1996. *Anniston Chemical Agent Disposal Facility Phase I Quantitative Risk Assessment*. Aberdeen Proving Ground, MD: US Army.

US Army. 1997. *Pine Bluff Chemical Agent Disposal Facility Phase I Quantitative Risk Assessment*. Aberdeen Proving Ground, MD: US Army.

US Bureau of the Census. Undated. *US Interim projections by Age, Sex, Race, and Hispanic Origin: 2000–2050*. www.census.gov/population/www/pop-profile/nat proj.html.

US Bureau of the Census. Undated. State and County QuickFacts. http://quickfacts. census.gov

US Bureau of Reclamation. 1980. *Animas-La Plata Project, Colorado–New Mexico, Final Environmental Statement*. Washington, DC: US Department of the Interior.

US Bureau of Reclamation. 1992. *Animas-La Plata Project, Colorado-New Mexico. Draft Supplement to the 1980 Final Environmental Statement*. Washington, DC: US Department of the Interior.

US Bureau of Reclamation. 1996. *Animas-La Plata Project, Colorado–New Mexico, Final Supplement to the Final Environmental Impact Statement*, Vols I and II and Technical Appendices. Washington, DC: US Department of the Interior.

US Bureau of Reclamation. 2000a. *Animas-La Plata Project, Colorado–New Mexico, Final Supplemental Environmental Impact Statement*, Vols 1, 2, 3a and 3b. Washington, DC: US Department of the Interior.

US Bureau of Reclamation. 2000b. *Record of Decision. Animas-La Plata Project, Colorado Ute Indian Water Rights Settlement Final Supplemental Environmental Impact Report*. Washington, DC: US Department of the Interior.

US Bureau of Reclamation. 2002. *Animas-La Plata Project, Ridges Basin Dam and Reservoir, Pre-Construction Facilities Reclamation, Final Environmental Assessment; Petroleum Product Spill Analysis, attachment April; Draft Biological Assessment*, April. Washington, DC: US Department of the Interior.

US Bureau of Reclamation. 2009. *Animas-La Plata Construction*. August. www. usbr.gov/uc/progact/animas/update.html. Accessed May 21, 2010.

US Department of Energy. 1994. *Final Supplemental Environmental Impact Statement, Defense Waste Processing Facility Defense Waste Processing Facility*, DOE/EIS-0082-S. Aiken, SC: Savannah River Operations Office.

US Department of Energy. 1995a. *Final Environmental Impact Statement, Waste Management, Savannah River Site*, DOE-EIS-217. Aiken, SC: Savannah River Operations Office.

US Department of Energy. 1995b. *Record of Decision for the Defense Waste Processing Facility Defense Waste Processing Facility Environmental Impact Statement*. Aiken, SC: Savannah River Operations Office. *Federal Register* 60(70), FR 18589, April 12.

US Department of Energy. 1997. *Supplemental Record of Decision; Savannah River Site Waste Management*. Aiken, SC: Savannah River Operations Office. *Federal Register* 62(96), FR 27241, May 19, 27241–27242.

US Department of Energy. 2009a. *Supplement Analysis, Interim Management of Nuclear Materials, Final Environmental Impact Statement*, DOE/EIS-0220-SA-01. http://nepa.energy.gov/documents/EIS-0220-SA-01.pdf

US Department of Energy. 2009b. *States with Renewable Portfolio Standards.* http://apps1.eere.energy.gov/states/maps/renewable_portfolio_states.cfm

US Department of Energy. 2010. *Savannah River Site, History Highlights.* www.srs.gov/general/about/history1.htm

US Department of the Interior. 2007. *Oil-Spill Risk Analysis: Gulf of Mexico Outer Continental Shelf (OCS) Lease Sales, Central Planning Area and Western Planning Area, 2007–2012, and Gulfwide OCS Program, 2007–2046,* OCS Report MMS 207-040. Washington, DC: Minerals and Management Service, Environmental Division.

US Department of the Interior. 2010. *Service, Stewardship and Integrity. Annual Performance Report, FY 2009.* www.doi.gov/pmb/ppp/upload/DOI-APR_FY09.pdf

US Department of Transportation. 2007. *Draft Scoping Document: Northern Branch Corridor, DEIS, Draft Environmental Impact Statement. Federal Transit Administration and New Jersey Transit.* www.northernbranchcorridor.com/docs/Northern%20Branch%20Final%20Scoping%20Doc%20031108.pdf

US Department of Transportation. 2008. *I-70 East Draft Environmental Impact Statement.* www.I-70east.com/reports.html

US EPA. 1999. *Consideration of Cumulative Impacts in EPA Review of NEPA Documents,* EPA 315-R-99-0021. Washington, DC: US Environmental Protection Agency, Office of Federal Activities.

US EPA. 2010a. *Safe Drinking Water Act.* Washington, DC: US Environmental Protection Agency. http://water.epa.gov/lawsregs/rulesregs/sdwa/index.cfm

US EPA. 2010b. *Federal Water Pollution Control Act Amendments of 1972.* Washington, DC: US Environmental Protection Agency. www.epa.gov/history/topics/fwpca

US EPA. 2010c. *CERCLA Overview.* Washington, DC: US Environmental Protection Agency. www.epa.gov/superfund/policy/cercla.htm

US EPA. 2010d. *Survey of the Toxic Substance Control Act.* Washington, DC: US Environmental Protection Agency. www.epa.gov/lawsregs/laws/tsca.html

US EPA. 2010e. *Summary of the Resource Conservation and Recovery Act.* Washington, DC: US Environmental Protection Agency. www.epa.gov/lawsregs/laws/rcra.html

US Fish and Wildlife Service. 2009. Hotchkiss National Fish Hatchery begins stocking fish for Animas-La Plata Project. www.fws.gov/mountain-prairie/pressrel/09-46.html

Vecsey G, 1971. Eastern Kentucky, scarred by strip mining, looks to T.V.A. suit. *The New York Times* March 7, p. 36.

Vicente G, Partidario M. 2006. SEA – enhancing communications for better environmental decisions. *Environmental Impact Assessment Review* 26: 696–706.

Volpe Center. 2010. *Visualization Case Studies: A Summary of Three Transportation Applications of Visualization.* Washington, DC: Planning and Policy Analysis Division, John A. Volpe National Transportation Systems Center/Federal Highway Administration, US Department of Transportation. www.gis.fhwa.dot.gov/documents/visual_toc.htm

Waggoner W. 1973a. Turnpike agency contesting Byrne. *The New York Times* December 19, p. 47.

Waggoner W. 1973b. Byrne says plan to add to turnpike are being postponed. *The New York Times* December 28, p. 61.

Walker S. 2005. *Shockwave.* NY: Harper Collins.

Warnback A, Hilding-Rydevik T. 2009. Cumulative effects in Swedish EIA practice – difficulties and obstacles. *Environmental Impact Assessment Review* 29: 107–115.

Warner M. 1973. *Environmental Impact Analysis: An Examination of Three Methodologies*. Springfield, VA: National Technical Information Service.

Weiland P. 1997. Amending the National Environmental Policy Act: federal environmental protection in the twenty-first century. *Land Use & Environmental Law* 12(2): 275–305.

Wellinghoff J. 2009. United States of America, Federal Energy Regulatory Commission, AES Sparrows Point, LNG, LLC; Mid-Atlantic Express, LLC. January 15.

Wells J, Robins M. 2006. *Communicating the Benefits of TOD*. Washington, DC: US EPA, 6, 23–39.

Wenner L. 1982. *The Environmental Decade in Court*. Bloomington, IN: Indiana University Press.

Wentworth S. 2007. O'Malley asks feds to deny Sparrows Point LNG project. *Baltimore Business Journal* February 7. http://baltimore.bizjournals.com/baltimore/stories/2007/02/05/daily28.html

WILPF. Undated. *Chemical Weapons*. Geneva: Women's International League for Peace and Freedom. www.reachingcriticalwill.org/legal/cw/cwindex.html

Winter JC, Ware JA, Arnold PJ III, eds. 1986. *The Cultural Resources of Ridges Basin and Upper Wildcat Canyon*. Albuquerque, NM: University of New Mexico, Office of Contract Archeology. (Submitted to US Department of the Interior, Bureau of Reclamation.)

Wood G. 2008. Threshold criteria for evaluating and communicating impact significance in environmental statements: "see no evil, hear no evil, speak no evil"? *Environmental Impact Assessment Review* 28: 22–38.

Wood P. 2003. FERC chairman welcomes fed chairman's focus on LNG. Press release, June 11. Washington, DC: Federal Energy Regulatory Commission.

Yoo J. 1998. The new sovereignty and the old constitution: the chemical weapons convention and the appointments clause. *Constitutional Commentary* (online subscription) 15: 87. http://web.Lexis-nexis.com/universe

Zukin C. 1999. A quarter-century of New Jersey development: successes and failures. *The Star-Ledger/Eagleton-Rutgers Poll*. Release SL/EP 74-2, December 19. http://slerp.rutgers.edu/retrieve.php?id=124-2

# Index

Note: Page numbers in *italics* refer to figures and those in **bold** type refer to tables

Chester, Pennsylvania 200
China 22
CIF *see* Consolidated Incineration
Facility (CIF)
Coast Guard 80, 82, 84, 93, 107
Colorado River Basin Project Act
1968 178
Colorado Ute Indian Water Rights
Settlement Act 174, 178
Comprehensive Environmental
Response Compensation Liability
Act (Superfund) 205
Connaughton, James 16
Conrail 40
Consolidated Incineration Facility
(CIF): Savannah River Site (SRS)
154, 164, 166–7
consultation: NEPA requirements 3;
*see also* scoping process;
stakeholders
cost-benefit analyses: Animas-La
Plata (ALP) project 177
Council on Environmental Quality
(CEQ): capacity 10; guidance on
climate change 208–9; guidelines
for preparation of EIS 4; *The
National Environmental Policy Act,
A study of its effectiveness after
twenty-five years* 19–20;
responsibilities 6; weaknesses 13
Coverdale & Colpitts, Inc. 34
cultural and historical attributes:
Ellis Island 54, 56, 59–60
cumulative impacts 15, 23, 51,
199–200; Animas-La Plata (ALP)
project 190; Ellis Island 76;
Johnston Island 126; Sparrows
Point liquefied natural gas (LNG)
EIS 99

Defense Waste Processing Facility
(DWPF): Savannah River Site 141,
143, 147, 152, 166, 170
Denver I-70 East corridor 201
Department of Energy (DOE) 139,
144, 159–61, 170; Durango
Uranium Mine Tailings Remedial
Action (UMTRA) 183;
environment of major sites 147;
site-specific advisory boards
(SSABs) 163, 169
Department of Transportation
(DOT): onshore LNG facilities 80,
82; visualizations 202

Dinar, A. *et al.* 174
draft EISs 5–6
Dreher, Robert 18–19
Dreyfus, D. and Ingram, H. 8, 9
DWPF *see* Defense Waste Processing
Facility (DWPF)

economic costs of EIS process 15–16,
19
ecosystems 147–8, 157–9; *see also*
aquatic ecosystems
E.I. DuPont de Nemours & Co. Inc.
145
electronic formats 201–2
Ellis Island Development Program:
economic resources 66–8, 75; EIS
elements 63, **64**; EIS evaluation
75–7, 198, 199–200, 204;
historical and cultural resources
64; history of the proposal 60–1;
natural resources 65; no-action
alternative 61–2, 64; preferred
alternatives 62–3, 64, 66–7; social
resources 65–6; stakeholder
reactions 68–70; value of the EIS
process 71–4
Ellis Island Institute 62, 71, 72, 73
Ellis Island, New York Harbor 54, **55**,
**57**; deteriorating structures 54,
56, *58*, 60, 61–2; immigration
function 54, 56, 59, 73;
sovereignty dispute 60
Ellison, B. 203
endangered species: Four Corners
178, 182, 185–6; Johnston Atoll
122–4, 126; Savannah River Site
(SRS) EIS 147, 158–9
Energy Information Administration
(EIA) 78
environmental assessments (EAs) 5,
6, 19–20, 171
Environmental Impact Statements
(EIS): changes in plans proposed
165; consciousness-raising 11;
covering multiple projects 168;
improvements needed 192;
initial use by environmental
organizations 10–11; legal
mandate 3, 4–5; process 5–7, 22;
programmatic 23; public-private
partnerships 73–4; regulatory
79–80, 91, 93–4; technical
complexity 160–1; *see also*
National Environmental Policy

120; *see also* Second
Supplemental Environmental
Impact Statement (SSEIS) at
Johnston Island
Joyner, Marsha 130

Kean, Thomas 40, 42
Kelliher, Joseph 103
Kohl, Helmut 127, 132
Kosson, David 133–4

Lawrence, D. 14
legal challenges: Johnston Atoll
132
Liberty State Park, New Jersey 65, 70
liquefied natural gas (LNG) 78;
demand and supply 106–7;
hazardous nature of 81;
importance for USA 78–9; tankers
83–5, *83*, 87, 93; US regulations
for facilities 80–2; *see also*
Sparrows Point liquefied natural
gas (LNG) EIS
local government EIS processes
12–13
low-level nuclear waste 144

McMorris, Cathy 16
Martin, John 19
Medvedev, Dmitry 142
Menendez, Robert 50, 70, 74–5
methane 81, 192
Mid-American Pipeline Corporation
(MAPCO) 181, 182, 190
Mid-Atlantic Express 83, 102
Mid-Atlantic Express Pipeline Project
80, *85*, 99; impacts on aquatic
systems 96, 97; socioeconomic
impacts 98; visual impacts 97
MILVANs 122, 128
Montoya, Stella 175–6
Mormons 174
Morris Udall Foundation 7
Muskie, Edmund 7, 8, 10

National Environmental Policy Act
(NEPA): architects of 7–9;
evaluation questions 21–3;
historical background 1; imitation
processes by states and other
nations 12–13, 22; impact on
New Jersey highways projects
29–30; initial interpretations 4–7;
initial media attention 9–11;

objectives 2; post-year-2000
assessments 16–21; role and
effectiveness 205–6; strengths
11–13; *Title 1, Sections 101–103*
2–3, 4–5, 8, 13; *Title 2, Section 201*
4; weaknesses 13–16; *see also*
Environmental Impact
Statements (EIS); planning
through the EIS process
National Environment Research
Parks 147
National Fire Protection
Organization (NFPO) 80
national parks: New York Harbor 54,
*55*, 56
National Park Service (NPS): Ellis
Island 56, 60–3, 66, 68–9, 71,
71–3; Governors Island 67, 76
National Technical Information
Service 11
natural gas 78, 192; pipeline
relocations *see* Animas-La Plata
(ALP) project, Final
Environmental Assessment; *see
also* liquefied natural gas (LNG)
Navajo Nation 173, 180, 193
NEPA *see* National Environmental
Policy Act (NEPA)
neutralization of chemical weapons
116
New Jersey: counties *26*; effects of
road building 27; Gold Coast *28*,
40; highway proposals and the
EIA process 29–30;
suburbanization and population
growth 25–6, 39; traffic
congestion 26–7; *see also* Alfred
E, Driscoll Expressway; Hudson-
Bergen Light Rail System
New Jersey Highway Authority
30
New Jersey Turnpike Authority 30–1,
32–3, 35, 45–6
New York City 206
New York Harbor: national parks 54,
*55*, 56
Nighthorse, Lake 178, *179*
Nixon, Richard 7–8, 10
NJ Transit (NJT) 39, 42–3, 44, 49
Northwest Pipeline Corporation 181,
182
nuclear power generation 143
nuclear waste 143–5; *see also*
Savannah River Site (SRS) EIS

Nuclear Waste Policy Act (NWPA) 1982 144
nuclear weapons 139, 141–2; production 145; waste management 140–1, 142–5

Obama, Barack 142
occupational exposures and safety: Johnston Island 124–6; Savannah River Site (SRS) EIS 161–3
O'Donnell, Lauren 104–6
oil spill analysis 206–8; pipe relocation FEA 184–5, 190

Pantex, Texas 149, 156
Person, J. 15
Pick-Sloan Plan 192
Pine Bluff weapons stockpile **114**, 115
pipelines *see* Animas-La Plata (ALP) project; Mid-Atlantic Express Pipeline Project
Pipeline Safety Act (1979) 80
planning through the EIS process 11, 13, 21, 195–6; communications 200–2; cumulative impacts information 199–200; Ellis Island 71, 72, 77; environmental justice and socioeconomic information 197–8, 200; flexibility 202–4; Gulf Coast oil platform failure 206–8; Hudson-Bergen Light Rail System 49–51; impact on global warming 208–9; regional economic input 197
plutonium 142, 143, 145; waste management 124, 155, 164, 165
policy-evaluation criteria 21–2
pollution: air 27; DOE major sites 147; metals contamination 49, 161; radioactive contamination 124, 147–8, 160–1; water 205–6
Popper, Frank 191–2
private sector development 14, 23
Proxmire, William 10
public participation 12; Animas-La Plata (ALP) project 189–91; Ellis Island Development Program 69, 76–7; Johnston Atoll EISs 129–32, 136; Northern Branch Corridor DEIS 46, 47, 49–50; role of FERC 106; Savannah River Site (SRS) EIS 163–5, 169–70; Sparrows Point EIS 101–2

race and ethnicity *see* immigration; Ute Indian tribes
radioactive contamination 124, 147–8, 160–1; worker and public health 161–3
Rahall, Nick 17
Reagan, Ronald 127, 132, 174
remediation: impacts on ecosystems 147–8
renewable energy 106–7
Resource Conservation and Recovery Act 1976 205
risk analyses 206–8, 209
Robins, Martin 42–3, 47, 49–51
Robinson, Mark 109
Rosen, Amy 43
Ruckelshaus, William 10
Ruppersberger, Dutch 103

Safe Drinking Water Act 205
Salazar, Ken 74–5
San Francisco 22
Savannah River Site (SRS) EIS 141, *146*, 149, *150*, *151*, 152; changes in plans 165; DWPF 143; ecological impacts 157–9; environment and long-term stewardship 145–8, 166–7; evaluation 167–70, 204; facilities operating beyond 2024 **157**; federal laws and regulations 143–5; four options offered 152, **153**, 154–6; groundwater impacts 160–1; impacts considered 156, **158**; land use impacts 157; public reactions 163–5, 169; socioeconomic impacts 159–60, 168, 197; worker and public health impacts 161–3
Save Ellis Island, Inc. (SEI) 63, 71–2, 73, 74
science-driven reform mechanism 20
scientific evaluation of impacts 14–15
scoping process 5, 12; Animas-La Plata (ALP) pipeline relocation 189; Ellis Island Development Program 68, 69; Hudson-Bergen Light Rail System 45–7, 51, 200; Johnston Island EISs 128; Sparrows Point liquefied natural gas (LNG) EIS 101–2
Second Supplemental Environmental Impact Statement

(SSEIS) at Johnston Island 117,
119, 121; accident analyses
125–6; alternative actions 127–8;
compliance with regulations
128–9; cumulative impact
assessments 126; environmental
impacts 122–4, **123**; ethical
dilemmas 121; evaluation 133,
134–8, 198, 203; preferred action
121–4; public response 129–32
socioeconomic impacts 197–8;
Savannah River Site (SRS) EIS
159–60, 168; Sparrows Point
liquefied natural gas (LNG) EIS
98–9
Sparrows Point liquefied natural gas
(LNG) EIS 79, 80, 81, 82–3, 88;
alternative actions 90–1; channel
dredging 87–9; cumulative
impacts 99, 199; DEIS conclusions
99–101; EIS evaluation 106–10,
197–8, 204; getting to the
unloading sites 83–5; impact on
aquatic systems 96–7, 99–100;
impacts considered **92–3**; location
84, 86; natural gas pipelines 85,
90; reliability and safety 91, 93–5,
99, 107–8; socioeconomic
impacts 98–9; unloading,
converting and sending out the
gas 89–90; vessel operation 87;
visual impact 97
Spaulding, Paul, III 132
stakeholders: Alfred E, Driscoll
Expressway 37–8; Animas-La Plata
(ALP) project 189–91, 193–4; Ellis
Island Development Program
68–70, 76–7; Hudson-Bergen
Light Rail System 52–3, 204;
Johnston Island chemical
weapons disposal 119, 129,
136–8; Savannah River Site (SRS)
EIS 163–5, 169–70; Sparrows
Point liquefied natural gas (LNG)
EIS 101–3, 108–9; *see also* scoping
process
*The Star-Ledger* 39
supersonic transport (SST) 10

thermonuclear weapons 142
Tooele weapons stockpile 113, **114**,
115
Toxic Substances Control Act 1976
205

Train, Russell 10
translations 201
transparency 200–2
transportation: chemical weapons
121–2, 133, 134–5; effects of road-
building 27; growth in
automobile use 25, 26–7; mass
transit 30; rejection of New Jersey
highways projects 29–30; SST 10;
*see also* Alfred E, Driscoll
Expressway; Hudson-Bergen Light
Rail System
transuranic wastes 143, 145, 155,
163, 164, 167
Tzoumis, K. and Feingold, L. 6, 7

Udall, Tom 17
uranium fuel 145
US Army *see* chemical weapons
Ute Indian tribes: relationship with
Anglo-Americans 175–6;
relocation to Four Corners area
171, 173–4, 193; spiritual
importance of water 174; water
rights 174, 177, 178

Van Dyke, Jon M. 132
Vicente, G. and Partidario, M. 15
visual aids 201–2

Waste Isolation Pilot Plant (WIPP),
New Mexico 143, 144, 152, 167
water quality 205–6; Sparrows Point
99, 103
water rights: Four Corners region
173, 174, 177, 178
wildlife: Animas-La Plata (ALP)
project 178, 180, 182, 185–6, 191;
exposure to nuclear materials
147–8; Johnston Atoll 122–4, 126,
131; New York Harbor 65;
pipeline and dredging impacts
96–7; Savannah River Site (SRS)
EIS 158–9
Wiygul, Robert 176–7, 194
Wood, G. 15
workers *see* occupational exposures
and safety
Wyoming 192

Yost, Nicholas 17–18
Yucca Mountain, Nevada 144